Revision  Guides

GCSE Geography

Liz Taylor

Illustrations by Matthew Foster-Smith

GCSE Geography

Name...

Class...

School...

....

Date and topics of exams:

1 ....

2 ....

3 ....

Exam board...

Syllabus number...

Candidate number...

Centre number...

Further copies of this publication may be obtained from:

Pearson Publishing
Chesterton Mill, French's Road, Cambridge CB4 3NP
Tel 01223 350555   Fax 01223 356484
Email info@pearson.co.uk   Web site http://www.pearson.co.uk/education/

ISBN: 1 85749 545 4

Published by Pearson Publishing 1998
© Pearson Publishing 1998

Contents

Introduction

Welcome to the **GCSE Geography Revision Guide**!

Here are a few pointers to help you get the most out of the book:

- **Use it alongside your class notes** – The book gives a good summary of key topics, skills and vocabulary, but it is not a substitute for two years' classwork, so keep your notes handy!

- The book covers the **main topics** for all GCSE syllabuses, but check with your teacher if any sections are not needed for your syllabus. For example, you may not need Hazards or Glaciation.

- **Case studies** are a really important part of GCSE Geography. Revise your key case studies by making summary cards and learning them (see page 2). Don't worry if you did different case studies to the ones in the book. If in doubt, use what you did in class or check with your teacher.

- **Use this book to test yourself** – Revise a section of your class notes then find the topic in the book by looking at the *Contents* on pages iii and iv. Read the appropriate page(s). Wherever there is a vertical line you can test yourself by covering the right-hand side with a piece of paper and writing your answers on the paper. It does not matter if your answers are not the same word for word so long as you have the right meaning. Test yourself on diagram labels in the same way.

 The **questions** throughout the book are also designed for you to test what you know, so revise the topics before having a go at them. If you write in the book, keep your writing small and neat. Answers to short questions are in the back of the book. If you are stuck, have a look then test yourself again the next day. It is always helpful to come back to a topic to check you still remember it.

 'Over to you' – These are longer questions for use after you have revised a topic. Check the answers in your notes, from a textbook or with your teacher. Try doing old exam questions too. Your teacher will have some, or you can order these from the exam boards.

Spider diagrams are included when a topic involves a lot of different factors. Write the key phrase in the middle of a rough notebook, then put the factors around the edge. For example:

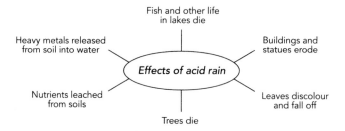

Key points are included in the answers section, but you may be able to think of extra ones.

 Words in **bold** are geographical terms. Check you can give a definition of these. Practise spelling these too! There's a list of useful spellings on page 17.

• Break up your revision into manageable chunks and use the **tick list** on the *Contents* pages to keep an eye on your progress. Don't leave everything until the last minute!

Note: MEDC = More Economically Developed Country
LEDC = Less Economically Developed Country

1 Skills

GCSE Geography tests skills as well as knowledge – all syllabuses will involve skills such as using maps and statistics.

Tips for technique in Geography exams

- Make sure that you know the dates and times of your exams. Some exam boards split the topics you have covered between the two exams, so it is important to know which topic will be in which exam. Some exams focus on your knowledge of the topics, so lots of revision is necessary. Some focus on applying skills and techniques you have learnt to new information. In this case, you may get some advance information about the topic, but the skills you have built up during the course will be of more help than last minute revision.

- Make sure you know the format of the exam before you go in. For example, will you have a choice of questions, or are all questions compulsory? If you have a choice, it is very important to read through all the questions before you choose. Be particularly careful over questions that ask you for a case study – these questions often have higher marks available, so make sure that the example you choose really fits the question.

- If you are instructed to write on the question paper, you are likely to need all the space provided for you. If you need more space, don't be afraid to continue onto blank areas of the page, or onto lined paper. If you use paper at the back of the answer booklet, make it clear to the examiner that the answer is continued later and number the question clearly on the continuation paper. If all your answers are on lined paper, it is a good idea to leave a few lines after longer answers just in case you want to go back to add something else.

- Look carefully at the number of marks allowed for the question before answering. This will usually be written near the margin, eg [6]. A one mark question usually expects a one word or one sentence answer. Two to six mark questions usually involve some level of explanation or reasoning, needing more than one sentence. A question over 6 marks usually expects an extended piece of writing or mini-essay. For longer questions, you may find it helpful to do a short plan or think through your main points before you start to write – try not to waffle.

- Watch your timings carefully and try not to spend longer than you should on one section of the paper. If you are getting stuck on a question, leave it and come back at the end. If you run out of time at the end of the exam, write short notes about what you would have said – a bullet point list is ideal. You will usually receive some credit for relevant answers in this form.

- Check whether one or both of your exams will involve a map question and make sure you are familiar with all the skills in this chapter.

- Don't be afraid to use sketch maps and diagrams in your answers – these are important parts of geography and can help you present information quickly (see page 5).

- Spelling, punctuation and grammar are assessed in Geography exams, so try to spend 5 minutes checking through your work at the end. The examiner will also be looking to see if you use geographical words correctly (eg 'urban', 'erosion').

Using and revising case studies

Case studies are real life examples. For example, a case study of a major shopping centre could be the MetroCentre, Gateshead. Using case studies well in the exam is important for getting the higher grades in both Foundation and Higher tiers.

You will have covered many case studies during your GCSE course. Some will be short examples and some will be long and detailed. It is important to work out which ones are most useful for the exam.

 Write down a list of topics you have studied (ask your teacher, or look at the syllabus) and write down your main case study for each topic, eg:

- Rural-urban migration: Belo Horizonte, Brazil
- Problems in cities in LEDCs: Nairobi, Kenya
- Solutions to urban problems: Nairobi, Kenya

You are likely to have between 10 and 20 major case studies to learn in depth, and some smaller examples.

- Some case studies can be used for more than one topic, eg the Amazon Rainforest, Brazil, can be used for structure and composition of an ecosystem, effects of people on an ecosystem, effects of people on the water cycle and ecosystem management.

- Remember to write the full location of the case study. For example, put 'Aswan Dam, River Nile, Egypt' rather than just 'Aswan Dam'.

- Any fieldwork you have done also counts as a case study. You may need to summarise what you have done. Ask to see your coursework again, if necessary.

 Try using index cards (available from newsagents) to summarise your case studies. Remember to include facts and figures.

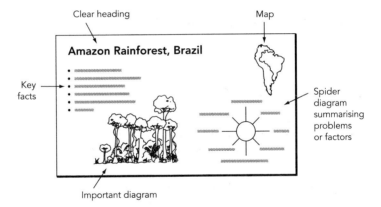

Clear heading — **Amazon Rainforest, Brazil**

Map

Key facts

Spider diagram summarising problems or factors

Important diagram

Decision-making exercises in exams

In decision-making papers, you usually answer all questions. The order of the questions usually follows the decision-making process below:

1 Describing information

Short questions, using OS maps, graphs or other sources.

2 Using the information

Perhaps working out different people's attitudes to a proposal or assessing the advantages and disadvantages of a scheme.

3 Making a decision

This may involve choosing a site or planning something. You may have to write down your plan or draw an annotated (labelled) map. There is usually more than one correct answer in a decision-making exercise. If you have thought through your choice carefully and looked at all the information and instructions, it should be acceptable.

4 Justifying your decision

Explaining why your choice or plan is the best. When you justify a choice, always say why you rejected the other options as well as why you did choose that particular one. This part of the question often has the most marks, so really explain it – don't just describe your plan again.

If the questions follow this pattern, try to answer them in order as you may need information that you look at in one question to answer the next one.

If you can, refer to similar real life examples that you have studied – but only if they are relevant!

Look at the questions carefully. Some candidates find it helpful to underline key words, eg 'Explain how your plan is best for the environment and for local people.' In this case, you should try to write roughly the same amount about the advantages of your plan for the environment and for local people – try not to rush one section.

If you are given the resources for the exam in advance, spend some time looking over them and highlighting key facts and patterns. Your teacher will help you in this.

 Find out whether your exam will include a decision-making exercise and, if so, what form it will take.

Drawing sketch maps and diagrams

There are two main ways that you are likely to use sketch maps and diagrams in an exam:

1 To back up a case study, eg 'Describe the land use pattern in a town you have studied. You may wish to include a sketch map in your answer' or 'With the help of a diagram, explain how relief rainfall occurs'.

2 A map of a scheme you design as part of a decision-making exercise, eg 'Draw an annotated map to show how the National Park should be managed'.

Maps and diagrams should be clear and simple. Remember that the geographical information is more important than how artistic your drawing is, though it should be neat.

it is best to use pencil for drawing and pen for labels. Coloured pencils are useful to make features clear (eg different types of roads, or to shade in areas being managed in different ways). Include clear labels on maps and diagrams and use a key when appropriate. Diagrams should be fully explained, and maps may be annotated (labelled) to show more information, eg reasons why a supermarket site was chosen (see the example on page 6).

Work out which are the main maps or diagrams you need to learn (your teacher will advise you). Draw a simple copy of each and look at it for a while, then cover it over and draw it on some rough paper, then check the original. Keep practising until it is correct, and remember to check it again before the exam. Remember that many of the labels will be things you already know!

Sketch map to show advantages of the location of a Tesco supermarket near Cambridge

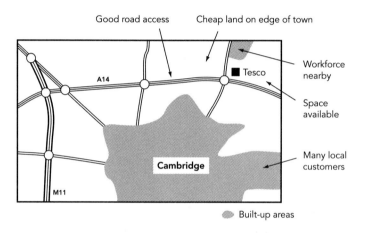

Map skills

Ordnance Survey (OS) maps are the standard maps for the UK. Most Geography exams will have at least one set of questions involving an OS map, most commonly at the 1:50,000 or 1:25,000 scale (see page 9).

Once you know the skills below, the best way of revising for map questions is getting used to looking at maps. If you don't have an OS map at home, there may be an extract in a textbook, or you could ask your teacher to let you borrow a map extract to help with your revision. Describe different areas on the map to yourself, and try to imagine what they would look like.

Symbols

Many map extracts used in GCSEs have a key (check this with your teacher), but it is a good idea to know the common symbols to save time and get a better 'feel' for the map. Note that the symbols on 1:50,000 and 1:25,000 maps are slightly different – usually in colour rather than shape.

 What do these symbols mean?

       [red dot]   [green]

 Now test yourself using the key on any OS map.

Compass directions

You will need to know the 16 point compass.

 Complete this one:

Remember that the top of an OS map is always North.

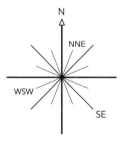

If you are asked to give compass directions, check whether the question says 'from' or 'to' a feature – it is easy to go the wrong way by mistake.

 Use the map below to fill in the compass directions on the table:

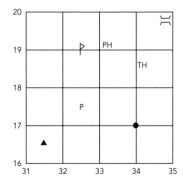

From	To	Direction
P	ⴲ	
P	▲	
●	ⴲ	
▲	⏜	
PH	TH	

Grid references

There are two sorts of grid reference:

- 4-figure grid references, eg 0234 for a whole square
- 6-figure grid references, eg 021347 show a point within the square.

In the exam, you may be asked to give a grid reference for a feature on the map (the question will say what type of reference) or you may be asked what feature is at a particular reference.

When referring to a map, try to give grid references, eg 'The relief is hilly and there are some cliffs, such as at 015728'.

4-figure references

Always give the number along the bottom first. Some people remember this by 'Along the corridor and up the stairs'.

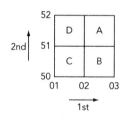

A = 02 51

B = _ _   _ _

C = _ _   _ _

D = _ _   _ _

6-figure references

Imagine each side of the square is divided into ten sections. If you want to be precise in an exam, each section will measure 2 mm on a 1:50,000 map.

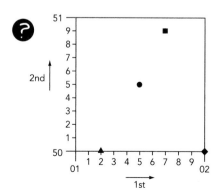
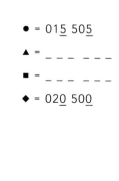

● = 015 505

▲ = _ _ _   _ _ _

■ = _ _ _   _ _ _

◆ = 020 500

Scale

OS map extracts usually include a scale at the bottom. On most maps, one grid square measures 1 km in real life, so you can work out the scale.

Scales may be given in three ways:

1   a ratio, eg 1:50,000 (1 cm on the map = 50,000 cm [$^1/_2$ km] in real life)

2   a statement, eg 2 cm = 1 km

3   a scale line

When measuring a straight line distance on the map, use a 30 cm ruler and read off the distance from the scale line, or calculate it, eg on a 1:50,000 map, halve the number of cm to get the distance in km, so 5 cm = 2 $^1/_2$ km. On a 1:25,000 map, divide the number of cm by 4 to get the distance in km, so 5 cm = 1 $^1/_4$ km.

If the scale is 1:50,000, how far is a measurement of:

1 cm? ....   4 cm? ....   12 cm? ....

If the scale is 1:25,000, how far is a measurement of:

1 cm? ....   10 cm? ....

If you are measuring a wavy line such as a road or river, you can use string (if you happen to have some!) or the edge of a piece of scrap paper (see below).

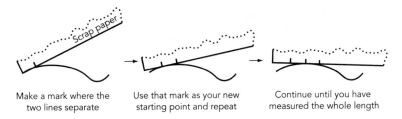

| Make a mark where the two lines separate | Use that mark as your new starting point and repeat | Continue until you have measured the whole length |

Try measuring this line! Scale: 2 cm = 1 km

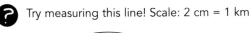

You may be asked to estimate the size of areas on the map. Often you will be given a choice of answers. Remember that 1 grid square = 1 km^2.

? Estimate the sizes of these areas:

Scale = 2 cm: 1 km

Area =

.... km^2

Area = ... km^2 Area = ... km^2

Height and relief

There are two main ways of showing height on OS maps:

1   Contour lines   —— 50 ——

2   Spot heights   • 57

All points along a contour line are at the same height. On 1:50,000 and 1:25,000 maps, the lines go up in tens, and the 50 and 100 metre lines are in bold.

On OS maps, the tops of the numbers always point upslope, so you should know which way is up, even if you can only see one number. (Note that this is not always the case on black and white maps in exam papers – use common sense.)

The closer the contour lines, the steeper the slope.

Steeper at top = Concave

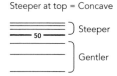

Steeper at base = Convex

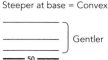

Look for rivers to help identify landscape shapes. Remember that they never flow uphill.

See page 12 for more practice in describing relief.

Statistical maps

Statistical information (eg percentage of people owning a car in an area) is often shown on maps because it is easier to see patterns. You should be able to interpret and add information to the following types of maps:

- **Choropleth** – shading shows values
- **Isopleth** – lines show values, like contours
- **Proportional symbols** – larger symbols (eg arrows, circles) show larger values. (Note that you will not be expected to draw proportional symbol maps in exams.)

 Fill in the blank area (Havering) with the corrrect shading for a -6.3% change.

- Describe the distribution of areas with -10% and over population change.
- What overall pattern does the map show?

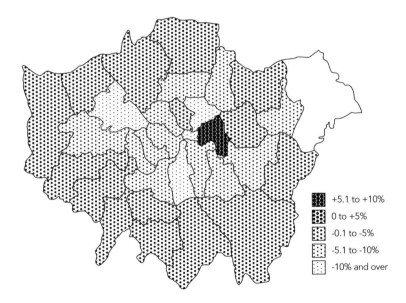

▦	+5.1 to +10%
▦	0 to +5%
▨	-0.1 to -5%
▨	-5.1 to -10%
▫	-10% and over

Sometimes exams include topological maps where relationships between places are correct but distances may be distorted, eg the London Underground map.

Describing maps

Both OS maps and statistical maps are common resources in Geography exams, and you could well be directly asked to describe what you see on them. The three questions below are frequently asked in exams:

1 Describe the relief of ...

Remember that relief is the shape of the land – just saying that it is high or low is not enough. Look at the map as a whole as well as different parts. Use grid references, accurate height measurements and compass directions, and name any landscape features you recognise.

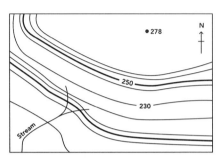

 Complete the description of the map above by selecting the correct words and filling in the gap.

The area is generally hilly/flat. It is higher in the north/south, reaching

a maximum of .... m. The slope is uneven/even, in a

concave/convex/stepped shape. The top of the hill is fairly wide and

flat, suggesting the landform may be a plateau/ridge.

2 Describe the location of ...

 Location means where something is.

Try to be precise and refer to nearby landmarks, distances and compass directions.

 Describe the location of Newtown Business Park.

The business park is located

.... km to the ....

.... of Newtown, on

the ....

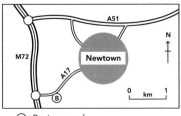

Ⓑ Business park

3 Describe the distribution of ...

 Distribution means how something is spread out over an area.

Look for patterns, use compass directions and count the features when appropriate.

 Describe the distribution of factories in Newtown.

All the factories are in the

.................................,
⸍
within .... km of the

CBD. 10 are to the

.... of the CBD

and .... are to the

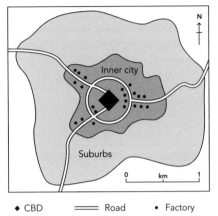

◆ CBD   ═══ Road   • Factory

east. There is evidence of

a .... pattern

along roads leading out of

the Central Business District (CBD).

 Useful words for describing distributions:

Linear	in a line.
Clustered, nucleated	in a clump.
Radial	coming out from the centre like spokes on a wheel.
Concentric	rings inside one another.
Random	no pattern.
Dispersed	spread out.

Graphs

There are many different types of graph which you may be expected to describe or add information to in your exam – check the exact requirements with your teacher. You may also be asked to calculate percentages or do other sums. Don't forget a calculator!

1 Bar graphs and line graphs

Some topics have specialised graphs, eg climate graphs (double axis – see page 116) and hydrographs (showing a river's discharge – see page 101).

If you are asked to add information to a partly-completed graph, try to continue by using the same type of symbols, line or shading pattern.

2 Stacked or compound bar graphs

Stacked bar graphs have more than one piece of information on each bar. Read off the bottom of the bar as normal, but then read the next section from where the bottom one ends – don't start at zero.

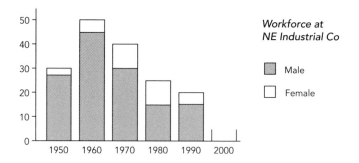

Workforce at NE Industrial Co

Male

Female

 Draw in the bar for the year 2000 – 20 men, 15 women.

How many workers were there in 1960?

How many were female?

Sometimes you will find just one large divided bar. These usually show percentages so the scale runs from 0 to 100. Again, each segment starts where the previous one ends, so when all the information is included, there should be no empty space.

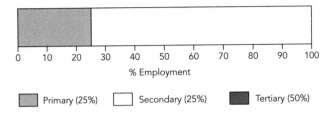

% Employment

Primary (25%) Secondary (25%) Tertiary (50%)

? Complete the divided bar above, showing the employment structure in Brazil (1996).

3 Triangular graphs

These are used to show data which is in percentage form and where each 100% is divided into three categories (eg employment structure). Each point on the graph adds up to 100. Use pencil guidelines if it helps.

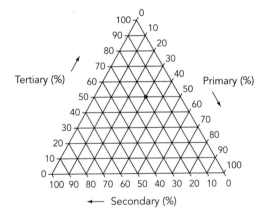

? Add the following countries to the graph:

* Bangladesh – 59% Primary, 13% Secondary, 28% Tertiary.
* USA – 3% Primary, 25% Secondary, 72% Tertiary.

4 Circular graphs

Rose diagrams and radial graphs are based on axes coming out from the centre of a circle:

* Rose diagrams are used to plot the number (or percentage) of something against compass directions.
* Radial graphs show the distance of things from a central place.

For how many days did the wind at Littlehampton blow from the East?

Rose graph showing wind direction at Littlehampton

 How many of the amenities shown below are 2 km or less from John's house?

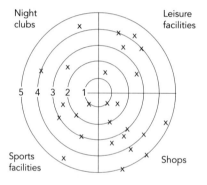

Radial graph showing distance from John's house to various amenities

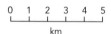

Centre point = John's house

5 Scatter graphs

These use a series of dots or crosses to show the relationship between two different sets of data. Sometimes a particular trend (pattern or direction) in the data will show up. For example, in the graph on the right, life expectancy increases with the wealth of a country (GNP). This is called a positive correlation. A negative correlation is the opposite (when one variable goes up, the other goes down).

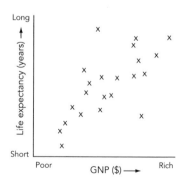

6 Pie charts

You may be asked to read figures off from a pie chart, or to complete one. If the figures are percentages, then multiply by 3.6 to get degrees. Remember to start each segment from where the previous one ends. Always start from 0 (at the top), with the largest section first.

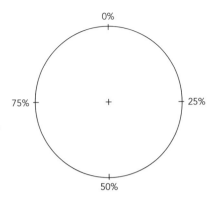

 Complete this pie chart to show tenure in East Devon:

Owner occupied housing 78%   78 x 3.6 = ... °

Private rented housing 12%   12 x 3.6 = ... °

Public rented housing 10%   10 x 3.6 = ... °

7 Describing graphs

Be detailed – look for trends and quote figures. The following is an example of a thorough description of a line graph showing London's population. 'London's population increased from 5.6 million in 1891 to a high of 8.6 million in 1939. It then declined steadily to 6.4 million in 1991.' Don't explain the patterns if you're only asked to describe.

Check your spelling!

Here are some words which are often spelt incorrectly in Geography exams. Look at the spellings, then get someone to test you on them.

- accessible
- affect (verb, eg how would the village be affected by…)
- congestion
- cramped
- deciduous
- development

- environment
- professional
- residential

- accommodation
- business

- corrugated iron
- dam
- decision
- effect (noun, eg the effects of the bypass are …)

- evaporation
- reservoir
- separate

2 Economic activities

This chapter covers different types of work, including farming, manufacturing and tourism.

Employment structure

 Employment opportunities (jobs) can be divided into three main groups:

Primary industry	obtaining raw materials, eg farming, fishing.
Secondary industry	manufacturing (making things), eg factory work.
Tertiary industry	services, eg shop work, teaching, healthcare.
Employment structure means	how employment in a place is divided between these three groups.
It also includes	how the jobs are split between full-time and part-time, male and female workers and different ethnic groups.
Quaternary industry is	high level service jobs such as research, information services, and administration.
Formal work is	officially paid and taxed, eg dentist.
Informal work is	unofficial and not taxed, eg baby-sitting.

Employment structures change over time

The graph shows how Britain's employment structure has changed since 1700.

Employment structure in the UK

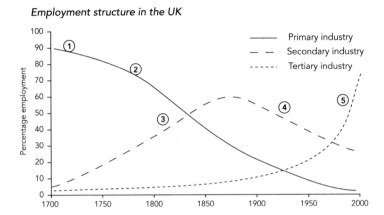

Explanation of the graph:

① Most people worked in	farming because most was done by hand.
② Mechanisation	machines were developed for farming, so fewer workers were needed.
③ Industrial Revolution	many people moved to the cities to work in factories.
④ Deindustrialisation	heavy industries (eg ship building) have declined due to changing markets and competition from abroad.
⑤ Tertiarisation	employment in services such as finance and tourism has increased.

? Describe the UK's employment structure in 1995.

Employment structure varies from place to place

LEDCs generally have more people employed in	primary industry.
For example, Ethiopia (1995)	P = 88%, S = 2%, T = 10%
MEDCs generally have more people employed in	tertiary industry.
For example, Japan (1995)	P = 7%, S = 34%, T = 59%

Triangular graphs are often used to show employment structure (see page 15).

The number of employment opportunities available also varies from place to place, eg in 1994, Spain had 24% unemployment whilst Luxembourg had 3%.

World trade

Trade is	buying and selling goods and services.
Imports are	goods brought into a country.
Exports are	goods taken out of a country.
A **trade surplus** is when	a country sells more than it buys, so it becomes richer.
A **trade deficit** is when	a country buys more than it sells, so it is short of money.
'Invisible' imports are	payments for services, eg money spent by foreign tourists or profit made on investments.

Patterns of trade

Many world trading patterns were established in colonial times when many European countries had Empires (eg India was a colony of the UK). The countries in Europe were able to get cheap raw materials from their colonies, which helped Europe become economically developed.

If an LEDC relies on one or two main products for export, what problems may it face? (For example, Angola gets 91% of its export income from oil.)

Trade groupings

 Trading blocs are | formed by countries joining together to look after their trade interests.

Examples include | the European Union (EU) or North American Free Trade Association (NAFTA).

They encourage | low cost trade between members by reducing or removing customs payments.

Countries outside the group are | charged higher customs payments (tariffs) to bring their products into the group. This makes goods from outside more expensive and so less competitive.

GATT stands for | the General Agreement on Tariffs and Trade.

It is aimed to | encourage free trade (removing tariffs) so all countries in the world could import and export their goods without paying extra.

The World Trade Organisation | replaced GATT in January 1995. It implements trade agreeements and attempts to settle disputes.

Farming (Agriculture)

Farms can be thought of as systems, with inputs and outputs.

 There are three main types of farm:

1 **Arable** means | growing crops.
2 **Pastoral** means | raising animals.
3 **Mixed** means | pastoral and arable on the same farm.

Commercial farming aims to	sell all or most of the produce to make a profit.
Subsistence farming is intended to	feed the farmer and his/her family.
Intensive farming requires a	high input of money or labour per km^2.
For example,	factory farming in UK or rice farming in Bangladesh.
Extensive farming involves	a lower input of money and labour per km^2 but each farm tends to be larger.
For example,	sheep farming in the Lake District.
Organic farming is	growing crops and rearing animals without using chemicals such as artificial fertilisers and insecticides.

 Draw a spider diagram of 'Factors affecting farm type'. One to start you off – 'Amount of rainfall'.

Farming in the UK

This simplified map shows the distribution of the main farm types in the UK:

 Think of five reasons why farm type is different between the west and east of the UK. Check your answer with your class notes.

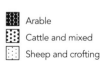

Arable

Cattle and mixed

Sheep and crofting

	Example 1: Hill farming	Example 2: Arable farming
Eg	Lake District or Welsh Mountains	East Anglia
Relief	Hilly or mountainous	Generally flat
Soil	Thin and poor	Deeper and more fertile
Climate	Harsh – long winters, short summers, lots of rain	Warm summers, relatively dry
Size of farms	Variable – from small family farms up to large estates (1000 hectares+) plus summer grazing on higher land.	Average 44 hectares, but many company-owned farms are much larger.
Producing	Sheep for wool and, on lower ground, cows for milk, both also sold to lowland farms for fattening for meat.	Crops – cereals (mainly wheat) and smaller amounts of other crops such as oil seed rape, sunflowers, and linseed. Some land now set-aside.
Income	Relatively low, often supplemented by EU subsidies and income from tourism, eg farmhouse bed and breakfast.	Generally fairly profitable, though EU subsidies for cereals are declining.
Labour	Small workforce, often just family members.	Small workforces due to a high level of mechanisation. Casual labour employed during harvest.
Capital input	Relatively low.	Large amounts of money tied up in machinery, also high expenditure on fertilisers and pesticides. Sometimes referred to as 'Agribusiness' because it is very commercial.

 Think of one way in which these two farming systems affect their local environment.

Farming in LEDCs

There is a wide variety of farm types in LEDCs:

Plantations are	large commercial farms in tropical areas, often growing just one crop for export, ie 'cash crops'.
For example,	tea in Sri Lanka.
Nomadic herding is	when people move from place to place to find grazing for their animals.
For example,	the Fulani in Africa.
Shifting cultivation is	when subsistence farmers living in the rainforest cut down a small area to grow crops and raise animals for a few years, then they move on and the natural vegetation is left to grow back.
For example,	the Amerindians in the Amazon Rainforest, Brazil.

 In your class notes, find your case study(ies) of farming in LEDCs. Summarise your notes using the same side-headings used for the UK case studies on page 23.

Changes in farming in the UK

 Draw a spider diagram of 'Changes in farming in the UK this century'. One to start you off – 'Hedgerow removal'.

When farmers were guaranteed good prices by the government for produce such as milk and grain, we produced more than Europe needed. This is called a **surplus**. The EU has changed its policies to try to reduce surpluses. Britain is trying three schemes in particular:

1 Set-aside gives farmers grants if	they take arable land out of food production.
2 Farm woodland scheme gives grants for	planting deciduous trees on arable land.
3 Farm diversification scheme gives money to	develop other land uses such as tourism and recreation (eg golf courses).

However, these schemes are only taking effect slowly, and environmental issues such as soil erosion, nitrate pollution of water and decline in wildlife numbers are serious concerns.

Nitrates come from	fertilisers applied to farmland.
They get into water by	run-off from fields and leaching.
Eutrophication is	the rapid growth of algae and weeds on streams and ponds, caused by the nutrient-rich water.
Water life is killed because	light is reduced and the amount of oxygen goes down.

 Why were over 25% of Britain's hedgerows removed between 1945 and 1990? What problems does hedgerow removal cause?

Farming change in LEDCs – The Green Revolution in SE Asia

Increased food supplies were needed because	there was rapid population growth in southeast Asia.
The Green Revolution led to	a big increase in rice production.
It involved	1 New high-yielding varieties of rice
	2 New technology, eg fertilisers, machinery.
It was most effective in	India and Pakistan.

 What were the disadvantages of the Green Revolution?

A smaller scale way of developing agriculture is to	use intermediate technology.
This is	encouraging the use of simple but effective equipment, training and farming methods, eg better hand tools or small scale hydroelectric power schemes.
For example,	training schemes for local vets in Kenya, using modern medicines and traditional plants.
Its advantages are	that it is cheaper, there is less to go wrong, it is better for the environment and encourages local skills rather than exports from MEDCs.
However,	because schemes are small-scale, it may take longer to have an impact over a whole country.

Manufacturing industry

Heavy industry is	manufacturing of heavy, bulky goods.
For example,	steel making or ship building.
Light industry is	manufacturing of smaller, light goods.
For example,	assembling sunglasses.
The **market** for a product is	the demand for it.
Capital means	money.
Labour means	workers.
Infrastructure is	the systems of services such as phone lines, roads, electricity.

Industrial location

Industries need to consider a wide range of factors when deciding where to locate.

Heavy industries are tied to	raw materials, eg iron ore
because	they are bulky and expensive to move.
Light industries are more	'footloose' – they have a wider choice of location
because	their raw materials or components are easy to transport.

 Draw a spider diagram showing 'Locational needs of a ship-building firm'. Think about the type of site a ship yard would need and things that should be close by. One to start you off – 'Suitable workforce nearby'.

The growth and decline of heavy industry – Iron and steel making in the UK

Before the eighteenth century	iron making was small-scale.
It used	local iron ore and charcoal for power.
In the eighteenth century	coke was used for power.
The factories were located	on coal fields, eg south Wales, West Midlands.
When local iron ore ran out	more was brought in from other parts of the UK.
Technology improved	and steel making was more efficient, but it needed fewer workers.
By 1950,	the UK's iron ore deposits were running out.
The iron ore now came from	abroad, eg Spain, Sweden.
So newer steelworks were located	on the coast.
Many steelworks closed because	it was hard to compete with other countries.
This led to	much unemployment – 195,000 jobs were lost between 1967 and 1992.

Study the maps below and describe how the distribution of integrated iron and steelworks changed between 1967 and 1997.

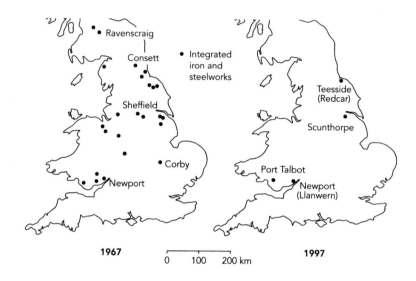

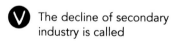

1967   0   100   200 km   1997

The growth of heavy industry during the industrial revolution had considerable effects on the environment, especially in the cities, which were growing rapidly. Air and water pollution were major problems. With the decline of heavy industry, industrial sites have had to be cleared and landscaped and any left-over land or water pollution problems dealt with.

V The decline of secondary industry is called **deindustrialisation.**

Draw a spider diagram showing 'Reasons why employment in heavy industry in the UK has declined'. One to start you off – 'World recession in 1970s'.

Case study:
Effects of deindustrialisation on the local area – Consett, Durham

Use the map (left) and an atlas, if you have one available, to describe the location of Consett.

Steel-making started in Consett in 1840. The town grew up around the factory and it was the main employer. The steelworks closed in 1980, with the following effects:

- nearly 4000 jobs lost from the steelworks

- unemployment rates up to 26% in 1982, male unemployment even higher

- associated industries, eg the coal mine and a ball-bearing manufacturers closed, losing more jobs

- derelict site had to be re-landscaped

- council lost £1.5 million in rates that the factory had paid each year, so it was hard to put money into the area

- social life lost, eg factory sports club

- people had less money to spend, so shops lost trade.

Reindustrialisation

There are many ways in which the government and local councils have tried to encourage new industries to set up in areas of high unemployment. For example, Enterprise Zones and Urban Development Corporations (see page 80).

 Brownfield sites are

areas where old industries have been knocked down so the space is available for redevelopment.

Industrial estates are

areas set aside for light industry and warehousing. They may be set up on brownfield sites or nearer the edges of a town.

Greenfield sites are

areas on the edges of towns which have never been built on before.

In Consett, the Derwentside Industrial Development Agency was set up to encourage new firms to come to the area. It helped create 4000 jobs by 1989. The largest new firm is Derwent Valley Foods (makers of the Phileas Fogg brand of snacks). The redevelopment had some success, but cost the council £1.8 million between 1978 and 1987.

 If you studied a different case study for growth and decline of an industry, write out a summary card for it, including:

- location(s)
- reasons why it grew up there
- facts about its growth
- reasons why it declined
- effects of the decline
- attempts at redevelopment (if appropriate).

High-tech industry

 High-tech industry is

the design and manufacture of technologically sophisticated products, eg microchips.

In the UK, high-tech grew rapidly during

the 1980s.

High-tech firms tend to cluster together, often on

science parks.

These are

landscaped areas set aside for high-tech firms, often with links to a university.

They are often located on

'greenfield sites' at the edge of towns.

The first science park in the UK was

Cambridge Science Park, set up in 1972.

Name the concentrations of high-tech industry on this map of the UK:

Suggest some reasons for the difference in location between R&D and manufacturing on the map.

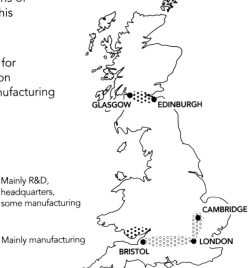

Mainly R&D, headquarters, some manufacturing

Mainly manufacturing

Factors affecting the location of high-tech industry

High-tech industry is **'footloose'**. This means that it doesn't have to be near its raw materials because the components used are usually relatively light and easy to transport so it has a wide choice of locations. Getting a skilled workforce is very important to a high-tech firm.

 Finish annotating this diagram with reasons why the Cambridge Science Park is in a good location.

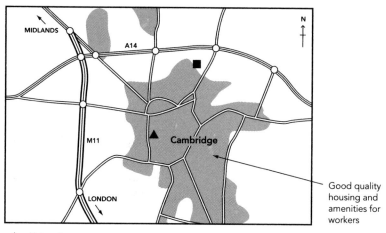

Good quality housing and amenities for workers

▲ University
■ Science Park
● Built-up areas

Once an industry has started to grow in an area, it tends to attract other industries. This brings money and people into the area. This knock-on effect is called the **multiplier effect**.

Add labels in the empty boxes on this diagram showing the multiplier effect:

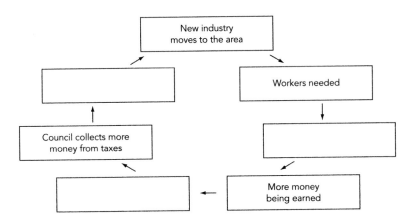

What may happen to the area if this cycle continues for a long time?

Industry in LEDCs

Some LEDCs have developed secondary industry a lot more than others.

NICs are	Newly Industrialising Countries – countries which have undergone rapid and successful industrial growth over the last 30 years.
They include	Taiwan, South Korea, Singapore and Hong Kong.
They were successful due to	initially cheap and plentiful labour supplies and investment from MEDCs, eg Japan.
Growth of industry brought	increased wealth, eg Taiwan's GNP per capita rose from $120 in 1960 to $11,000 in 1993.
A disadvantage was	high pollution due to lack of controls (eg Taiwan).

The workforces in NICs are now	more skilled and more expensive.
So they are locating factories in	countries with cheaper labour forces such as Indonesia.
Multinational companies are	large firms with factories in more than one country.
For example,	Nissan, Cadbury-Schweppes.
They are also called	transnational companies.
Their head offices are usually in	MEDCs, eg Nissan in Japan.
Most manufacturing is done in	LEDCs
because	production costs are cheaper.

 What are the advantages and disadvantages of locating a branch factory in an LEDC for:

a   the multinational company?

b   the country where the branch factory is located?

 From your class notes, write out a summary card for a case study of a multinational company. Include:

- some facts about the company (eg size, date set up, products, turnover)
- locations of branch plants
- how it has affected one area where it has a branch plant.

Tourism

Tourism is classed as	tertiary industry.
International tourism developed from	the 1950s onwards.
Now international tourists make	almost 500 million trips per year.
Most tourists are from	MEDCs.

Tourism is the world's fastest growing industry because	1 many people's incomes have increased
	2 people in MEDCs have increased leisure time due to early retirement and longer holiday allowances
	3 travel has become faster, easier and more comfortable – increased car ownership, more extensive motorway networks, etc.
Tourism provides	employment and income
but income from tourism can vary	as fashions change.
In Britain, employment is often seasonal, which means	more people employed in summer than in winter.
Many jobs are	part-time and have a fairly low level of pay, eg waiting on tables, selling ice cream.

An example of:

• A traditional English seaside resort is	Blackpool, Lancashire.
• A popular 'sunshine' holiday area in the EU is	Costa del Sol, South Spain.
• A winter sports area in the EU is	Chamonix, French Alps.
• A centre for 'cultural' sight-seeing is	Rome, Italy.
• An LEDC promoting tourism is	South Korea.
• An LEDC experiencing mass tourism is	Kenya (safaris and coast).
• A country developing eco-tourism (sustainable tourism) is	Belize, Middle America.

Note: There are many other examples – don't worry if yours are different!

Effects of tourism on the environment – Lake District, UK

The Lake District is one of the UK's 12 National Parks.

 Fill in the names of the National Parks on the map in the key below:

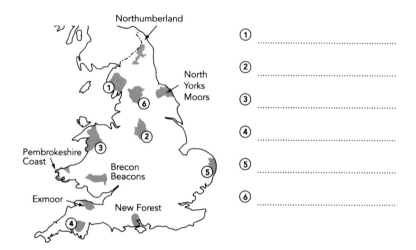

① ....

② ....

③ ....

④ ....

⑤ ....

⑥ ....

The Lake District receives	approximately 12 million visitors each year.
They are attracted by	pleasant scenery, opportunities for walking, water sports, sight-seeing (eg Wordsworth's cottage).
Some towns are particularly popular, eg	Windermere, Ambleside.

'Honeypot sites' are | areas which attract very large numbers of tourist for their size.

The honeypots gain from
- jobs in tourism, especially in the summer
- better incomes for shops and other services.

But have problems with
- congestion and air pollution from traffic fumes
- shortages of parking space
- footpath erosion and erosion on lake edges
- overcrowding, litter
- second homes left empty over the winter
- disturbance to farmland and wildlife.

Possible solutions include
- charging tolls on roads over the summer
- closing some roads, using park and ride buses
- much higher parking charges
- footpath management and paving.

There may be a conflict of interests between | tourism and other land uses, such as farming, which also need to use the area – 59% of land in the Lake District is owned by private landlords.

 Draw a spider diagram showing 'Possible advantages and disadvantages of tourism in an LEDC'. An idea to start you off – 'Attracts more people to the area'.

3 Ecosystems & environmental issues

This chapter is concerned with the environment – how it works and how it is affected by people. It includes a section on energy production.

How ecosystems work

 An **ecosystem** is

a community of living things and their non-living environment. The different parts of an ecosystem are linked together.

An example of an ecosystem is

a wood or a pond.

Living things are called

biotic, eg a cat.

Non-living things are called

abiotic, eg the sun.

Parts of the ecosystem are interdependent – this means

they need each other.

For example, plants need sunlight to

photosynthesise (produce energy)

Vegetation means

plants.

Cycles in ecosystems

 Look at the diagram and fill in the missing words in the passage on the next page:

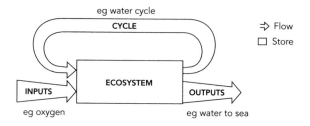

Materials and energy go into an ecosystem, travel through it and go out again. Things going in are called .... Examples include .... and oxygen. When something is travelling, it is called a ...., eg water in a river. If something stays in the ecosystem for a time, it is called a ...., eg carbon in a tree. Things leaving the system are called ...., eg oxygen produced by plants. Energy and materials travel round in a series of loops called .... , for example ....……… . Processes may change things, for example the process of evaporation is when water changes from a .... into a ....

The diagram below shows the carbon cycle. Fill in the missing labels. (If you get stuck, use your class notes or your textbook.)

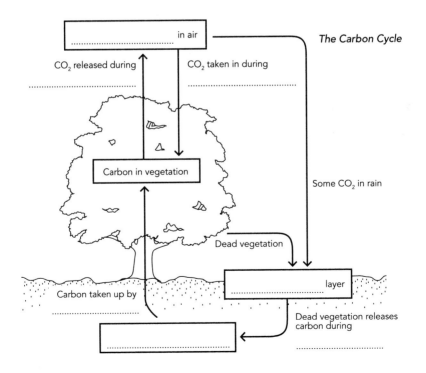

The Carbon Cycle

in air

CO₂ released during

CO₂ taken in during

Carbon in vegetation

Some CO₂ in rain

Dead vegetation

layer

Carbon taken up by

Dead vegetation releases carbon during

The carbon cycle is part of | the nutrient cycle.

Nutrients are | the food that plants need, eg nitrogen, magnesium.

Another cycle is | the water (hydrological) cycle (see page 94).

The energy for these cycles comes from | the sun.

Tip

Sometimes nutrient cycles are shown by diagrams where the symbols for the stores (usually a circle) and flows (arrows) are proportional to the amount of nutrients stored or flowing. So the wider the arrow, the more nutrients in the flow.

Food webs

 A **food chain** shows | how energy moves through an ecosystem by one member eating another.

The arrows point towards | the things that do the eating, eg cat ◄——— mouse.

A **food web** is | a more complicated diagram showing a series of food chains which are interlinked.

At each stage in the food chain, energy is | lost through processes such as respiration (breathing).

This means that higher up the food chain, | there are fewer animals.

? Study this diagram of a salt marsh ecosystem and fill in the table below:

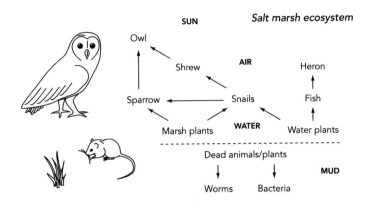

Term	Meaning	Example
Producer	Produces its own energy from inorganic materials	
Primary consumer		
Secondary consumer		
Herbivore		
Carnivore		
Decomposer		

? On the salt marsh diagram, underline those parts of the ecosystem which are biotic in one colour and abiotic in another.

Biomes

 A **biome** is   a large-scale ecosystem.

For example,   the Amazon Rainforest, Brazil.

The main factor controlling
the distribution of biomes is   climate.

They are also affected by   relief and soil type.

Many biomes have been
changed by   people, eg cutting down the
rainforests.

Biomass means   the amount of living things in an area
(plants and animals).

It is measured in   grams per square metre (g/m^2).

 Fill in the key to identify the four biomes on the map below:

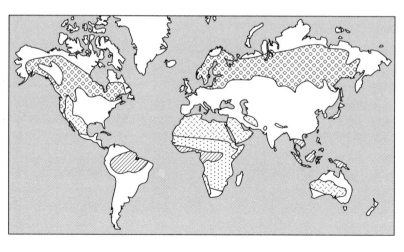

Characteristics of biomes

This table summarises the characteristics of desert and taiga (coniferous forest) ecosystems. Test yourself on it by covering over the second and third columns and jotting down what you can remember for each side heading.

	Coniferous forest (taiga)	Hot desert
Distribution	Generally in North – Canada, Scandinavia, Russia.	Generally between 15-30° north or south of the equator. For example, Sahara (Africa), Atacama (South America).
Climate	Cold (down to -30°C in winter) and dry (less than 500 mm precipitation per year).	Hot (temperatures up to 58°C, wide range between day and night) and dry (arid), though heavy rain does fall occasionally.
Soil type	Podsol	Very little soil, hardly any organic matter.
Vegetation	Coniferous trees, eg pine, spruce. Not much ground vegetation.	Sparse except after rain. For example, cacti, thorns.
Growing season	Relatively short: 6-8 months	Theoretically all year, but growth is slow due to aridity.
How vegetation is adapted to climate	Trees thin to shed snow, mostly evergreen to get maximum benefit from the growing season.	Plants grow and flower very quickly after rain. Cacti store water. Many plants are thorny to deter animals.
Wildlife	Relatively sparse. For example, owls, wolves, marmots, small birds.	Relatively sparse. For example, insects, snakes, gerbils.
Biomass	Average of 20,000 g/m²	Average of 700 g/m²
Impact of people	Cutting down trees for timber. Some pollution from oil drilling and pipelines in Siberia. Effects of acid rain or developing the area for skiing.	Relatively low. Some irrigation to make farmland. Damage to Arabian Desert from 1990 Gulf War.

Tropical rainforests

Tropical rainforests are located	between 5° north and south of the equator.
Temperatures are	high and relatively constant (25-30°C).
Precipitation is	high – over 2000 mm per year.
Rain usually falls as	convectional rain – thunderstorms on most days.
This climate is ideal for	plant growth.
So biomass is	high – 45,000 g/m².

 Biodiversity means — the variation in types of living things.

Biodiversity in the rainforest is — high – in the Amazon there may be more than 300 species of trees per km².

Wildlife is — abundant and diverse – 10 km² may contain 400 species of bird and 100 species of reptile.

 Label the diagram showing the different layers of rainforest vegetation.

Vegetation layers in a rainforest

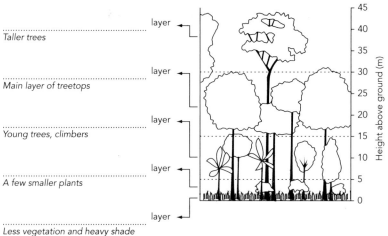

.... layer
Taller trees

.... layer
Main layer of treetops

.... layer
Young trees, climbers

.... layer
A few smaller plants

.... layer
Less vegetation and heavy shade

How is vegetation in the rainforest adapted to the climate?

The growing season is	all year, though individual trees lose their leaves from time to time.
Nutrient cycling in the rainforest is very	rapid.
Dead matter decomposes quickly due to	the hot, wet conditions.
Most nutrients are stored in	the biomass.
Few nutrients are stored in	the soil.
The soil type is	tropical latosol.

Deforestation in the rainforest

 Deforestation means the removal of forest cover.

The amount of rainforest cut down each year is around	40 million acres.
The most rapid loss is in	Brazil.

 Draw a spider diagram showing 'Reasons why the rainforest is being cut down'. One to start you off – 'Grazing land for cattle'.

Case study: Deforestation in the Amazon Rainforest, Brazil

Before the 1950s, the Amazon Rainforest supported an abundant plant and animal life plus around 1 million indigenous people. The people used the forest's resources in a sustainable way. From the 1950s, companies started to clear the forest to find valuable hardwoods like mahogany. Some trees were also cut down for paper, though the mills now use plantations of eucalyptus. More recently, agriculture has been a threat – 2.5 million hectares have been burnt down to give land for grazing beef cattle, but land is only fertile in the short term. Other areas of forest have been lost through government road-building programmes (eg the Trans-Amazon Highway). The government was trying to encourage people to move to this 'undeveloped' area. Some people moved in from poor areas of Brazil, such as the north-east, but this was not successful as the soil was not fertile enough for settled farming. Mining projects were more profitable, but involved clearing forest, eg at Trombetas.

Effects of deforestation

The rainforest ecosystem is very fragile. Once burned or cut down, trees take time to regrow and the ecosystem may never regain the diversity of life it once had. Deforestation has much wider effects than the local area.

 Draw a spider diagram showing 'Effects of cutting down the rainforest'. One to start you off – 'Extinction of plants which may have medicinal value.'

 Explain how deforestation can be the cause of:

a   flooding

b   disease amongst indigenous peoples

c   reduced rainfall in the area.

Forest management

Resources are	things we can use, eg wood, coal, water.
Exploitation means	using a resource, eg the forest. (It can also mean using something in an unjust way.)
Conservation means	looking after something – stopping it being destroyed.
Sustainable management is	using something in a careful way so that it will last in the long term.
A **biosphere reserve** is	a large area where the natural ecosystem is in good condition and which is protected by the government.
An example is	Korup Forest Reserve in Cameroon, Africa.

This diagram shows the typical zones in a biosphere reserve:

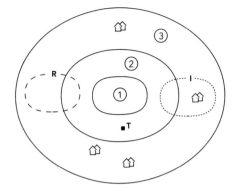

① Core area
just conservation

② Buffer zone
human presence controlled

③ Outer zone
suitable development allowed

∎T   Ecotourism base

– – R   Research

······ I   Area for indigenous peoples

 Settlements

Reserves may be linked by	wildlife corridors to enable migration.
The reserve is managed by	local people, with the help of conservation bodies.
The parks may make money from	ecotourism or selling local products.
Finance may also come from	'Debt-for-Nature' swaps – MEDCs pay off some of the debts of an LEDC if the LEDC spends money on conservation.
This has happened in	Bolivia and Ecuador, South America.

Ⓥ Agroforestry is — planting crops in between the trees, eg coffee in Colombia.

It is sustainable because — the nutrient cycle is not interrupted.

Trade in hardwood can be managed by — labelling wood that comes from sustainable sources and preventing illegal logging of hardwood trees.

Savannas

Savannas are located	in tropical areas, often in the area between the rainforest and the desert – mostly 5-15° north and south of the equator, eg North Australia.
Temperatures are	high (average 25°C).
Rainfall is	seasonal – up to 80% of rain may fall in four or five months. Total rainfall ranges from 500 to 2000 mm, depending on the location.
Much water is lost through	evapotranspiration from plants.
The soil type is	ferruginous.
'Closed' savanna is	mainly trees, with areas of grass.
'Open' savanna is	mainly grassland, with a few trees.
The growing season is	4 to 11 months, depending on location.

This diagram shows typical vegetation in the savanna:

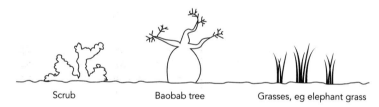

| Scrub | Baobab tree | Grasses, eg elephant grass |

Savanna vegetation

 How is savanna vegetation adapted to the climate and frequent fires?

Average biomass in the savanna is	4000 g/m^2.
The savanna has over 40 different species of large herbivores, eg	wildebeest, zebra.
There are fewer carnivores, eg	lions.
A major threat to the savanna is	desertification.

Desertification

Desertification is	the process whereby once productive land becomes desert.
Areas most at risk are	places with a long dry season, eg savannas and areas with a Mediterranean type climate.
For example,	savanna in the Sahel region of Africa (south of the Sahara Desert).
Natural factors involved are	drought and fires, but vegetation can usually recover from these in time.

 Draw a spider diagram showing 'Ways people may cause desertification'. One to start you off – 'Overcultivation – using the land too intensively for crops'.

Careful land management can reduce the threat of desertification. For example:

• Afforestation –	planting trees – holds soil together to reduce erosion, encourages infiltration, creates a windbreak, provides a managed source of firewood.
• Terracing –	making hillsides into large 'steps' which can be farmed – this reduces run-off and soil erosion. Lines of stones are used in the same way.
• Contour ploughing –	ploughing from side to side on a slope rather than up and down – reduces soil erosion.
• Improving soil with manure –	adds organic matter to help retain water, makes soil more fertile.
An example of management is	the Shewatta Project, Ethiopia.

The effects of energy production

Types of energy

 Draw a spider diagram showing 'Types of power supply'.
One to start you off – 'Oil'.

Fossil fuels are	the fossilised remains of plants and animals.
The major fossil fuels are	coal, oil and gas.
A **finite** resource is	one that will run out if we keep using it, eg oil.
An **infinite** resource is	one that will not run out, eg solar power.
Renewable energy is	energy that will not run out, eg wind power, or that regenerates and will not run out if carefully managed, eg wood.
Non-renewable energy is	a power source that can only be used once, eg oil. It does not regenerate within human time scales.
Energy use is higher in	MEDCs, eg in 1992, North America used 351 gigajoules of energy per person, compared to 36 gigajoules in Africa and the Middle East.
Energy efficiency means	not wasting energy, eg using energy-saving light bulbs.
A country's **energy mix** is	the combination of different energy sources it uses.

Describe the UK's energy mix.
Suggest how the energy mix
for an LEDC might be different.

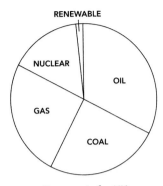

Energy mix for UK

Both renewable and non-renewable energy
sources have some effect on the environment.
However, the effects of renewable energy
sources tend to be much more localised,
eg noise from wind turbines. Burning fossil
fuels produces air pollution which may affect
a wide area, and is responsible for acid rain
and contributes to global warming.

 Make a table comparing the advantages and disadvantages of one
renewable and one non-renewable energy source you have studied.

Acid rain

Acid rain is | rain with a pH of less than 5.

 Complete this diagram to show the causes of acid rain:

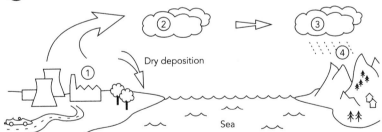

① .... and .... gases are released when

.... and .... are burnt in

..........................., ... and ....

② .... occurs in clouds and acids dissolve in moisture.

③ Clouds carried by ....

④ The most acidic rain recorded was pH .... in ....

Draw a spider diagram showing 'Effects of acid rain'. One to start you off – 'Erodes buildings'.

Areas badly affected by acid rain include	Scandinavia, Germany, East USA and Canada.

Solutions to acid rain

The effects of acid rain can be reduced by	adding lime to acid lakes and the areas surrounding them.
This works because	lime is an alkali, which counteracts the acid.
Disadvantages are	it is expensive and temporary.
An example of a lake which has been limed is	Loch Fleet, Galloway, Scotland.
The causes of acid rain can be reduced by	preventing emissions of sulphur dioxide and nitrogen oxide, especially from power stations.
This can be done by	technology such as fuel desulphurisation.
Alternatively,	the use of renewable energy can be encouraged.

Global warming

Global warming is a rise in average world temperatures. Temperatures change naturally in cycles which include the Ice Ages, but scientists think that current temperature rises may be caused by people. It is hard to estimate how much temperatures may rise because we do not fully understand how the Earth's systems work.

The **greenhouse effect** is	the ability of gases to insulate the Earth, keeping some of the sun's heat in – like glass in a greenhouse.
Greenhouse gases include	carbon dioxide, CFCs, methane and nitrous oxide.

| We need greenhouse gases because | without them, Earth would be 33°C colder. |

 In the space below, draw a diagram to show how the greenhouse effect works.

| |
| |

People may be increasing the level of greenhouse gases by	• burning fossil fuels • burning forests • leaving waste to decompose in landfill sites • releasing CFCs in aerosols and foam.
Global average temperatures may	rise by 1°C by 2030 and 3°C by 2100. Other estimates are higher.
This may cause average sea levels to rise due to	sea water expanding as it warms and ice melting.
This would particularly affect	low lying countries, eg Bangladesh, Pacific Islands, Netherlands.

 Draw a spider diagram showing 'Future effects of global warming'. One to start you off – 'Farming areas shift position as climate alters'.

Solutions to global warming

Demand for energy is likely to continue growing as the world's population grows and people expect a better standard of living. However, it is clearly necessary to reduce emissions of greenhouse gases.

This could be done by

- encouraging use of natural gas, which is cleaner than oil and coal (but natural gas supplies are finite)

- continuing to develop nuclear energy (though this may cause other problems)

- encouraging use of renewable energy such as solar and wave energy

- encouraging energy efficiency, eg energy-saving light bulbs

- using less intensive farming methods and stopping deforestation

- encouraging the use of public transport, walking and cycling

- investigating cleaner forms of power for transport, eg solar power.

International agreements on reducing CO_2 emissions (eg 1992 Rio Conference) are important, but countries are not always good at keeping to their targets.

4 Population & development

This chapter is about how the amount of people varies from place to place. It also covers quality of life.

Global population distribution

 Population means · the total number of people in a place.

Population density is · the number of people per km².

High population density means · lots of people in the area – densely populated.

For example, · Hong Kong has a population density of 6061 people per km².

Low population density means · few people in the area – sparsely populated.

For example, · Australia has an average population density of two people per km², because large areas of Australia are unpopulated, so these balance out the cities when the average density is calculated.

Most populations are not · evenly distributed.

Look at the map below showing world population distribution.

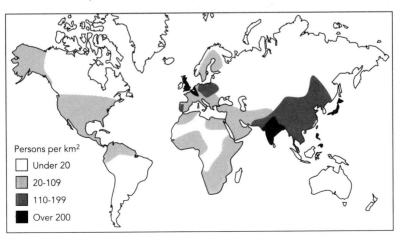

Persons per km²
- ☐ Under 20
- 20-109
- 110-199
- ■ Over 200

Name two areas on the map which are sparsely populated:

....

Name two areas which are densely populated:

....

Population density may also vary considerably within a country. For example, in Japan, the majority of the population lives near the coast as it is mountainous inland, whilst in Egypt, population density is very high next to the River Nile.

 Draw a spider diagram showing 'Factors affecting population density in an area'. One to start you off – 'What the climate is like'. If you get stuck, think about why lots of people live in London, or why no-one lives at the North Pole!

Population change

The population of an area rarely stays the same for any length of time.

The **birth rate** is	the number of babies born for each 1000 of the population in a year.
The **death rate** is	the number of people who die in each 1000 of the population in a year.
Infant mortality rate is	the number of babies which die before their first birthday, per 1000 live births in a year.
The **natural increase rate** is	the number of births minus the number of deaths per 1000 people in a year.

 Finish the sentences.

If birth rate is higher than death rate, the population will ....

If death rate is higher than birth rate, the population will ....

If birth rate is the same as death rate, the population will

....

The demographic transition model

The model below is a simplified version of how birth and death rates have changed over time:

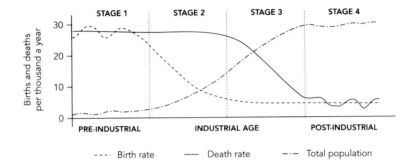

 Study the model and cross out the words which do not apply:

To start with, the world's population was small. Birth rates and death

rates were both high/low, so population change was fast/slow. As the

Industrial Age started, the birth rate/death rate fell, but the

birth rate/death rate remained high. This caused rapid growth/decline

in the total population. Eventually, birth rates increased/decreased, so

population growth increased/stabilised. In the post-industrial age,

birth rates and death rates are both high/low, which means the rate of

population change will be fast/slow.

Most MEDCs are now in stage 4 of the model, whilst LEDCs are mainly in stages 2 and 3. The model seems to describe adequately what has happened to the population in MEDCs like the UK, but it is a generalised picture and may not fit the history of any one country exactly. It may not predict the future for all LEDCs either because their economic situation is not the same as that of Europe in the Industrial Revolution.

World population

80% of the world's population live in	LEDCs.
In these areas, the growth rate is	high.
For example, Ethiopia's growth rate is	+29 per 1000 people (1991).
20% of the world's population live in	MEDCs.
In these areas, the growth rate is	low or stable.
For example, UK's growth rate is	+3 per 1000 people (1991).
The world's population is	rapidly increasing.
In 1995, it was estimated to be	5.7 billion.
Growth has been particularly rapid since	the 1950s.

Case study: Effects of rapid population growth

Most of the world's population growth is happening in LEDCs and often where there is a big gap between rich and poor. Population growth in most LEDCs is not happening at the same time as rapid industrial growth, as it was in most MEDCs, so there are fewer jobs available. Many people consider areas within LEDCs to be **overpopulated**. This means that an area has too many people for its resources to support. How you see population growth depends on your own values and attitudes. Some people feel that it is vital that LEDCs reduce their population growth rates because overpopulation may cause pollution and shortage of resources. Overpopulation in the Sahel is one cause of desertification due to people felling trees for firewood. MEDCs can cope better with high populations (eg in Tokyo land has been reclaimed from the sea for new housing). Other people think that there are enough resources in the world for everyone, but the problem is that they are distributed unfairly. The majority of the world's resource consumption and pollution is from MEDCs (eg MEDCs produced 95% of the global increase in CO_2 emissions which contribute to global warming).

 How can a government affect the population growth rates of its country?

Population structure

 **Population pyramids**
show

how the population of an area is divided up between male and female, and the different age groups. They can show figures for a whole country or just a particular town or rural area.

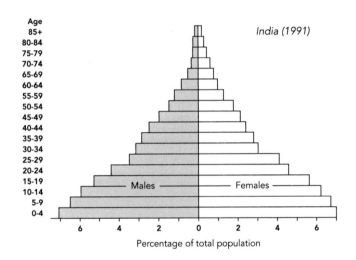

What percentage of India's population are men aged 45-49? ....

What percentage of India's population is aged 10-14? ....

What shape would the UK's population pyramid be?

....

Why is it different from India's?

....

Elderly dependants are

people aged 65+. The majority of this group do not work.

Young dependants are

children aged up to 15, the majority of whom are supported by their families.

Use the figures from the population pyramid to fill in this table:

	UK	India
Elderly dependants (%)	11.9	
Young dependants (%)	20.2	
Total dependants (%)		

 List the problems **and** benefits a country may experience from having:
a   a high percentage of elderly people
b   a high percentage of children.

Migration

Migration is — the movement of people from one place to another, with the intention of staying for more than one year.

Voluntary migration is — when people want to move, eg moving to be in an area with good sports facilities.

Forced migration is — when circumstances or other people make someone move, eg earthquake or war.

A **refugee** is — someone who has to migrate because they were in danger where they used to live.

Immigration means — moving into a country.

Emigration means — moving out of a country.

Recently, the majority of international migration has been	forced.
For example,	500,000 refugees moved from Rwanda into Uganda due to civil war.
Many migrants move from LEDCs to MEDCs to	find work.
For example,	4 million Mexicans crossed into the USA between 1980 and 1990. Much of this migration was illegal.
A **multicultural society** is	a place where the population has a mix of ethnic origins, eg UK.

 Find your case study of migration from your class notes. Fill out this factfile for it:

Name of country/area people moved from:	
Name of country/area people moved to:	
Number of people that moved:	
Why they moved:	
Advantages/disadvantages for the migrants:	
Advantages/disadvantages for the country they move **to**:	
Advantages/disadvantages for the country they move **from**:	

Why do many countries control immigration?

What types of people are most likely to be accepted?

Rural-urban migration

 Rural-urban migration is

	when people move from the countryside to live in urban areas. (Also known as rural-to-urban migration.)
This is a common trend in	LEDCs, eg Kenya, Africa.
Movement from urban to rural areas is called	**counter-urbanisation**.
This has mainly occurred in	MEDCs, eg UK. However, it has started to happen in some LEDCs too, eg from crowded Brazilian cities such as Rio de Janeiro.
Pull factors are	positive reasons why migrants want to move **to** the new area, eg more jobs available.
Push factors are	negative reasons why migrants want to move **away from** the old area, eg famine.
Depopulation means	when the number of people in an area falls. For example, the population of the Isle of Skye, UK, halved between 1901 and 1981.

 If rural-urban migration occurred on a large scale, how might the rural area be affected?

(Note: See *Chapter 5* for more information on the effects of migration on settlements.)

Development

Quality of life

 Quality of life means | how satisfied someone is with their living conditions and lifestyle.

We weigh up a huge range of things when assessing our quality of life. For example, our family circumstances, work, leisure, friends, environment.

 Draw a spider diagram showing 'Things that affect quality of life'. One to start you off – 'How good your health is'.

It is hard to assess quality of life because different things are important to different people. For example, one person may be content in circumstances another would find unbearable. Some aspects of quality of life are easy to measure, eg amount of living space can be measured in m^2 per person, but others are impossible to measure, eg happiness in family situation. The aspects of quality of life covered in the exams tend to be those which are easier to measure, eg income, environmental quality, healthcare provision.

Environmental quality means | how good an area is.

It could be measured by | the size and state of housing, amount of open space, range and size of amenities, etc.

On a global scale, the level of development of different countries is often compared.

 An **MEDC** is | a More Economically Developed Country.

This means | a richer country, eg Switzerland.

An **LEDC** is | a Less Economically Developed Country.

This means | a poorer country, eg Kenya.

The average quality of life is better in | an MEDC than an LEDC.

But it is important to remember | that quality of life varies in all countries – there will be rich people in an LEDC and poor people in an MEDC.

These are some common ways of measuring development:

• **Gross National Product**	(GNP) the total output of a country's economy in a year – measured in US dollars. Very basically, it means the wealth of a country.
• **Infant Mortality Rate**	number of babies which die before their first birthday – usually measured per 1000 live births.
• **Life expectancy**	how long, on average, a person can expect to live, measured in years.
• **Urban population**	percentage of the population who live in towns or cities.
• **Adult literacy**	percentage of adults who can read and write.
• Access to healthcare	measured as the number of people per doctor.
• Employment in services	(employment in tertiary sector) – measured as a percentage of total employment.
• Energy consumption	the average amount of energy used per person per year – often measured in kg of coal equivalent.

 Say whether each of the development indicators above would be high or low in a rich country.

 If a figure is shown as **'per capita'**, it means   for each person (on average).

Sometimes, a selection of indicators are combined to make   a quality of life index, eg the Human Development Index used by the United Nations.

The 'North-South Divide'

When levels of economic development are compared across the world, richer countries generally tend to be in the north and poorer ones in the south. The main exceptions are Australia and New Zealand, which are MEDCs but in the south. The term 'North-South divide' comes from a famous report about global inequality published in 1980.

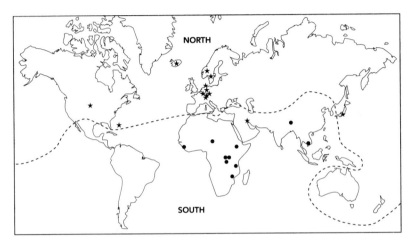

★ Richest MEDCs (GNP more than $24,000 per capita)

● Poorest LEDCs (GNP less than $200 per capita)

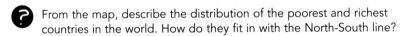 From the map, describe the distribution of the poorest and richest countries in the world. How do they fit in with the North-South line?

Remember that there isn't really a big 'gap' between MEDCs and LEDCs. There are also countries in the middle ground, eg Argentina (GNP per capita $7290 in 1993) and Portugal (GNP per capita $7890 in 1993). However, the difference between the richest and poorest countries seems to be increasing – the difference between the average income in the richest 20% and poorest 20% of countries grew from $1864 in 1960 to $15,149 in 1989.

Why is development unequal?

The reasons for some countries being more developed than others are very complex. Any or all of the following may be involved:

• Environmental factors –	suitable climate, soil fertility, ease of access.
• Population size –	under or overpopulation.
• Resources and technology –	availability of raw materials and developments in transport and research.
• History –	role of colonialism and relations with other countries.
• Political situation –	type of government, its aims and stability.
• Current investment –	by other countries and MNCs (multinational corporations) can seriously affect a country's economy by investing in or leaving an area. This is sometimes referred to as 'neo-colonialism'. Many poorer countries are in debt to MEDCs.

It is important to avoid oversimplifying these factors. For example, Japan has very little flat land and few raw materials, yet it is now one of the most economically developed countries in the world. It is also important to remember that a country's situation can improve or get worse over time.

Using your class notes, learn some facts and figures relating to quality of life and development for an MEDC and LEDC that you have studied. For example, you could learn their GNP, life expectancy and adult literacy figures. Check the exact requirements with your teacher or syllabus.

Regional differences in development

Development statistics tend to be averages or totals for a whole country. This can hide important differences between the regions. For example, in Italy, the northern region is considerably richer than the south.

Case study: Regional differences in Italy

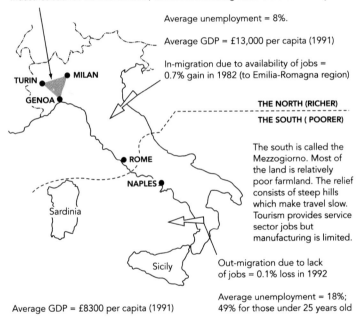

The 'Industrial triangle' – Three major cities with good communications and trade links. Industries such as car manufacture, fashion and banking make the area wealthy.

Average unemployment = 8%.

Average GDP = £13,000 per capita (1991)

In-migration due to availability of jobs = 0.7% gain in 1982 (to Emilia-Romagna region)

TURIN   ● MILAN

GENOA ●

THE NORTH (RICHER)

THE SOUTH (POORER)

● ROME

NAPLES ●

Sardinia

The south is called the Mezzogiorno. Most of the land is relatively poor farmland. The relief consists of steep hills which make travel slow. Tourism provides service sector jobs but manufacturing is limited.

Out-migration due to lack of jobs = 0.1% loss in 1992

Sicily

Average unemployment = 18%; 49% for those under 25 years old

Average GDP = £8300 per capita (1991)

If you studied regional differences in another country, draw your own sketch map and annotate it.

How do governments attempt to reduce regional inequalities?

Aid

 Aid is

any kind of help given from one country to another without direct payment being expected (though there may be other conditions involved).

 Draw a spider diagram showing 'Types of aid'. One to start you off – 'Training in healthcare'.

Short term aid is

given to solve immediate problems (eg food aid for an area with a famine).

Long term aid is

intended to have a lasting effect on quality of life in the area. It often involves teaching people skills so that they are more independent (eg training in preventing soil erosion).

 Colour-code your spider diagram to show short term and long term aid.

Aid may be given by

non-governmental organisations (eg Oxfam), governments or international agencies (eg the World Bank).

Bilateral aid is

direct transfer of aid from one country to another.

Multilateral aid is

transfer of aid from a number of countries through an agency such as the World Bank.

If aid plans are not carefully thought out, preferably involving local people in the decision-making, they can cause harm. For example, large-scale hydro-electric power plants will provide electricity but may force people to move when a valley is flooded to make the reservoir.

Projects involving western agricultural techniques may improve crop yields but make local people dependent on supplies of petrol and chemicals which are expensive and may harm the environment. Sometimes short term aid is vital, but the most effective aid is sustainable.

Case study of sustainable aid:
Upesi stoves in Kenya

'Intermediate Technology' is a charity which specialises in sustainable aid with maximum involvement of local people.

In Kenya, they work with local women's groups in rural areas to encourage the production and use of 'Upesi' stoves. These are pottery cylinders built into a mud surround which replace the traditional open fires. Their advantages are that they burn fuel more efficiently (conserving wood supplies and reducing the time that women spend collecting firewood), produce less smoke (reducing health risks) and can be made locally (providing steady employment).

5 Settlement

This chapter covers rural and urban settlements across the world and considers some of the issues that residents face.

Classifying settlements

(V) A **settlement** is | a place where people live.

Rural means | countryside.

Urban means | a built-up area (town or city).

A **hierarchy** is | when things are put into order according to their size or importance.

(?) Complete this table of a settlement hierarchy and give an example of a place you know for each level:

	Settlement type	Example
LARGEST	Megalopolis	Boston; New York; Washington, USA
↑	Conurbation	
	Town	
SMALLEST	Single houses	Remote farmhouses

A conurbation is	a city which has grown outwards and joined up with nearby settlements.
Rural settlement types include	villages and hamlets.
Urban settlement types include	towns and cities.

It is hard to give exact boundaries between large rural and small urban settlements. The following characteristics are usually taken into account:

Urban settlements have	• higher housing densities
	• larger populations
	• a greater number and variety of services
	• lower proportion of the population employed in primary sector jobs, like farming (though in many MEDCs, employment in primary jobs is now low in rural areas too).

Location of settlements

Many reasons for the locations of settlements are historical (eg good defensive positions) though the initial reasons may become less important over time.

 Draw a spider diagram showing 'Factors affecting the location of settlements'. One to start you off – 'Availability of water supply'.

V **Site** means	the land on which a settlement is built.
Situation means	the area around the settlement.

The diagram below shows the advantages of the site and situation for one Cambridgeshire village:

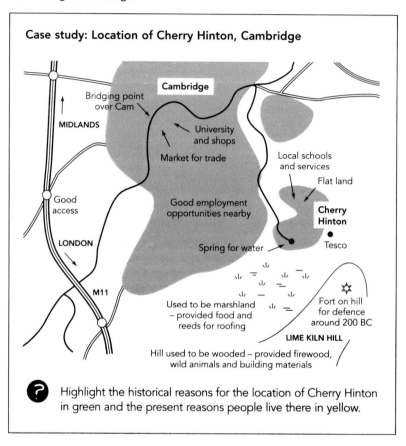

Case study: Location of Cherry Hinton, Cambridge

Highlight the historical reasons for the location of Cherry Hinton in green and the present reasons people live there in yellow.

An example of:

a town built inside a meander for defence is	Durham
a town on a bridging point across a river is	Cambridge
a town built on a higher point to avoid marshland is	Ely
a town built on a high point for defence is	Edinburgh

Settlement functions

 Function means the purpose of a place. A settlement's main economic activity is often referred to as its function, eg a tourist town or a fishing village.

Most urban settlements have more than one function. For example, London is a centre for finance and administration as well as a major tourist destination. Functions may change over time. For example, in many coastal towns, employment in tourism is now more important than fishing.

 List the functions of the settlement where you live or go to school.

Settlement patterns

Settlement patterns vary from area to area depending on factors such as relief, drainage and transport links.

 Give the correct name for these three settlement patterns:

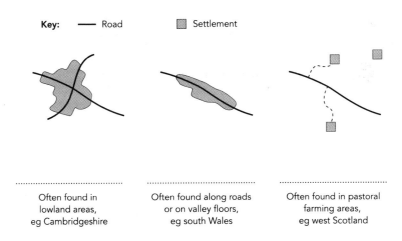

| Often found in lowland areas, eg Cambridgeshire | Often found along roads or on valley floors, eg south Wales | Often found in pastoral farming areas, eg west Scotland |

Global urbanisation patterns

 Urbanisation means | an increase in the proportion of the population living in towns and cities.

The 'level of urbanisation' in a country refers to | the percentage of the population who live in urban areas.

Levels of urbanisation in MEDCs are | generally high, eg UK 89%, Japan 77%.

Levels of urbanisation in LEDCs are | usually lower, but rapidly increasing, eg Brazil 76% (36% in 1950), Kenya 25%.

'**Millionaire**' or '**Million Plus**' cities are | cities which have over 1 million inhabitants.

There are 276 millionaire cities in the world (early 1990s figures). The map below shows the cities with over 5 million inhabitants:

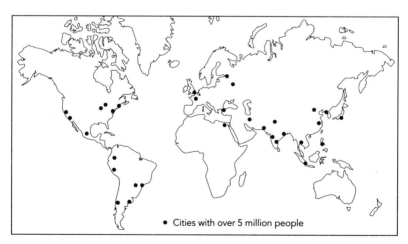

• Cities with over 5 million people

 Describe the distribution of the cities on the map.

What proportion of the cities with a 5 million+ population are located in MEDCs? (Tip: Check back to page 65 for a map of MEDCs/LEDCs.)

Cities grow by	in-migration and natural increase.
Cities are growing faster in LEDCs due to	high rural-urban migration and rapid natural increase.

Urban settlements

Land use patterns in MEDCs

Although all cities are different, most include the same types of land use.

 Draw a spider diagram showing 'Types of land use in a city'. One to start you off – 'Open space'.

Cities in MEDCs often share similar characteristics because they were shaped by similar processes of growth and change. Cities often have areas of one particular type of land use – these are called land use zones. For example, a zone of low cost housing. There are various urban land use models. Burgess based his model on a pattern of concentric rings – a simplified version is shown below:

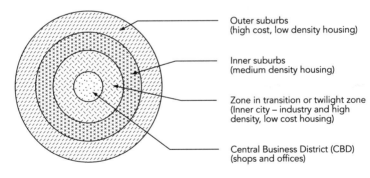

Outer suburbs
(high cost, low density housing)

Inner suburbs
(medium density housing)

Zone in transition or twilight zone
(Inner city – industry and high density, low cost housing)

Central Business District (CBD)
(shops and offices)

The model does not take into account some features of modern cities such as	• edge of town shopping areas • edge of town council estates • redevelopment in central areas.
An alternative model is	Hoyt's sector model.
Instead of rings, this model has	sectors or wedges of land use coming out from the centre along transport routes.

Fill in the blanks:

Quality of life and environmental quality vary across the city. Generally the

……………............. is thought of as having the lowest quality of life. This is

because the housing is ……….. there (often Victorian terraces) and there is

less open space. However, many inner city areas have now been

redeveloped (eg London Docklands), which may ………….. quality of life.

Population density is generally ……….... in the inner city and decreases

towards the …………… of the town. The inner suburbs are often made up

of ……………… housing built in the …………. . The outer suburbs have

………….. housing and modern estates. Although in Burgess' model,

industry is near the ……………, in many towns it has now moved outwards

to ………………….............................. .

 Check your notes for a case study of land use patterns in a city or
town. Draw yourself a simple copy of the land use pattern and learn it.

Land use patterns in LEDCs

The same types of land use are present in cities in MEDCs and LEDCs, but
the patterns tend to be different. Of course, there is also a lot of variation
from one LEDC to another. For example, African cities tend to have
different patterns to South American cities. The model below shows a
typical land use pattern in a Brazilian city:

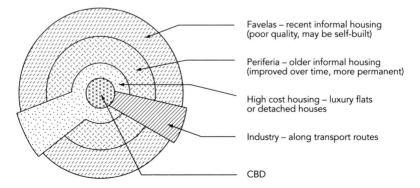

Favelas – recent informal housing
(poor quality, may be self-built)

Periferia – older informal housing
(improved over time, more permanent)

High cost housing – luxury flats
or detached houses

Industry – along transport routes

CBD

Factors such as relief will affect the shape of the city – a perfect circle is unlikely! For example, Rio de Janeiro in Brazil, is on the coast and the high class housing is concentrated along the coast due to the pleasant environment. The shanty towns (called favelas in Brazil) spread inland from the CBD and rich areas. 17% of the population live in favelas. The largest is Roçinha (80,000 people).

Outward growth of cities

Most cities grow outwards over time – even when the total population of the city may be falling, as in some cities in MEDCs.

In MEDCs, outward growth is called	**suburbanisation.**
It consists of	mainly good quality housing, often semi-detached or detached, sometimes in estates.
The housing density is	low.
If the growth is widespread and uncontrolled it is called	urban sprawl.
For example,	Los Angeles, USA, has vast areas of low density suburban housing.
One way of controlling urban sprawl is	to set up a '**green belt**'.
This is	a ring of land around a town or city in which planning permission is not usually granted for urban land uses such as housing.
A **New Town** is	a new area of housing and industry outside the green belt to take overspill population from the older city.
For example,	Stevenage is a New Town for London.
In cities in LEDCs, urban sprawl is caused by	rapid population growth and rural-urban migration.
The housing is	high density and poor quality (see page 75).

Urban issues – MEDCs

Access to housing

 Tenure means the way in which a home is owned. The three main types of tenure are:

- **owner occupied** (the person living there has bought the property)
- **privately rented** (rented from an individual or firm which owns the property)
- **local authority rented** (rented from the council).

Housing associations also provide rented housing – often to particular groups, eg elderly people. People do not have equal access to all housing types.

A **mortgage** is a loan from a bank or building society to enable someone to buy a home.

A person is usually allowed to borrow around three times their annual income.

A person is more likely to rent a home than to buy if their income is too low to buy or if they are wanting somewhere for a short time.

If a person's income is low, they may be eligible for local authority rented housing.

This is allocated on a points system according to need and how long they have been waiting.

 What types of housing are available in an area you know? How good is access to housing for people on a low income?

Change in the inner city in the UK

The **inner city** is	an area of housing and some industry next to the CBD.
In Victorian times, inner cities consisted of	factories and terraced housing for the workers, plus local services such as shops and churches.
For example,	the Lower Don Valley in Sheffield.
In the twentieth century, much of the industry has	closed down or moved to the edges of towns.

This led to various problems in the inner cities.

 Draw a spider diagram showing 'Inner city problems'. One to start you off – 'Unemployment'.

Following World War II, much inner city housing was	demolished.
This was called	**slum clearance**.
The inhabitants were moved to	high rise flats in the inner city, or edge of town estates.
This was not fully successful because	communities were broken up and some new buildings were made of poor materials which caused problems such as dampness. Tower blocks were awkward for young families and elderly people.

Case study: Inner city redevelopment in Glasgow, Scotland

Glasgow's inner city grew up to house workers in heavy industry such as ship-building on the River Clyde. Many workers lived in stone-built flats called tenements. When the industries declined, unemployment was high (male unemployment was 23% in 1971) and the housing quality was poor. The tenements did not have running water or inside toilet facilities. Between 1957–1974, the council tried to improve the inner city through **comprehensive redevelopment** in areas such as Gorbals. This involved demolishing large areas of tenements and replacing them with high-rise flats. Some people were also rehoused in council estates on the edge of town. An urban motorway was built to attempt to improve traffic conditions. The new flats were not popular and there were many problems in the area. More recently (1976-1987), the council has tried **renovation**. This is when the shells of existing tenements have been kept, but the insides have been fully modernised. 1200 homes were modernised and some new housing has also been built. This scheme was called the GEAR project (Glasgow Eastern Area Renewal). Renovation has been more popular with residents, and some jobs have been attracted to the area, but there are still some problems.

Other government initiatives to solve inner city problems include:

- Urban Development Corporations — organisations appointed by the government to oversee redevelopment. They concentrate on improving the environment and **infrastructure** and encouraging private investment.

  For example, the London Docklands Development Corporation.

- Enterprise Zones — small areas of land where the government offer **incentives** such as reduced taxes to encourage businesses to locate there.

  For example, the Isle of Dogs Enterprise Zone, London.

Infrastructure means main services like water, electricity, roads and communications.

Not all redevelopment is organised by the government:

(V) Gentrification is | when wealthier people move into inner city areas and renovate the housing.

They want to be near the CBD | because it is closer for work and has good facilities.

An example is | Islington, London.

Access to services

A person's income and where they live affects their access to services such as entertainment, sports, shops and education.

There is a hierarchy of service provision with larger settlements having a greater number and greater variety of services.

High order services are | specialised services, eg a wedding dress shop.

They are usually found in | larger settlements such as towns and cities.

Low order services are | more general services, eg grocer shop.

They are found in | most types of settlements.

There is also a hierarchy in shopping provision within an urban settlement.

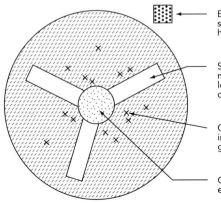

Edge-of-town shopping mall or superstores selling a variety of high and low order goods.

Shopping streets usually along main roads running into town – low order food shops and higher order, eg kitchenware.

Corner shops usually located in inner city areas, selling low order goods. Many now closed down.

CBD generally higher order shops, eg clothes. May include shopping malls.

Choose **two** of the shopping areas on the map on page 81 and explain:

a why the locations are good for those types of shops

b how growth or decline in those types of shops would affect local people.

| An example of a large, out-of-town shopping centre is | the MetroCentre, Gateshead. |

Urban transport

Urban areas across the world are experiencing problems handling an ever-increasing volume of traffic. Trends such as counter-urbanisation and the development of out-of-town shopping centres rely on high car ownership, but most cities were not planned with this in mind.

 Draw a spider diagram showing 'Urban transport problems'. One to start you off – 'Increased wear and tear on roads'.

 For an urban area you know or have studied, describe any solutions to transport problems being used (eg bypass, park and ride scheme) and evaluate how successful these have been.

Urban issues – LEDCs

Housing problems

All urban areas contain inequalities, but the rich-poor gap tends to be most obvious in cities in LEDCs. Most problems in these cities are caused by rapid, unplanned growth. The lack of formal sector jobs for newcomers to the city results in poverty, and local authorities do not have the resources for unemployment benefit.

Low incomes mean that private rented housing is often too expensive, so squatter settlements grow up on marginal areas of land – by rivers, roads, railways or on the edges of the city. For example, in Addis Ababa, Ethiopia, 85% of residents live in 'informal settlements'. These shanty towns suffer from lack of services and poor quality housing. Before the 1970s, many local authorities had refused to support them, often bulldozing them in the hope that the problem would go away.

For a squatter settlement you have studied, write a list of problems faced by the residents. Try to include some facts and figures.

Solutions to housing problems

Cover over the right-hand column and explain what these solutions involve. Give an example for each one:

Site and services scheme –

Housing scheme in which residents rent or buy a small plot of land with connection to mains services. They then build their own house, sometimes with the help of a loan for materials. Eg Dandora, Nairobi, Kenya.

Upgrading –

The local authority gives secure tenure to inhabitants and provides additional services such as street lighting, mains drainage or electricity to an existing informal housing area. Eg George, Lusaka, Zambia.

Relocation –

Residents in squatter settlements are moved to new housing, often large block of flats. Eg Singapore.

Building new towns –

New towns are built to reduce the number of people moving to the city. Eg Brasilia, Brazil or New Bombay, India.

Improving rural areas –

Villages are given better services and schools to try to stop the rural-urban migration. Eg Tanzania.

Rural settlements in MEDCs

Since the 1960s, there has been a trend of population movement from large urban areas to smaller towns and rural areas in MEDCs. This is called **counter-urbanisation**. From 1981-1991, large cities in the north of Britain lost 4% of their population, but rural areas in the north gained 5%.

Push factors from the cities include	• high levels of air pollution
	• concerns about crime
	• lack of open space.
Pull factors to rural areas include	• cheaper housing compared to nearby cities
	• lifestyle perceived to be more relaxed
	• more attractive environment.
This move was possible because	increasing car ownership makes the population more mobile.
It has led to an increase in	commuting.
Commuter villages are	villages where a large proportion of the population do not work in the settlement.
They are also called	dormitory or suburbanised villages.
For example,	Fulbourn, Cambridge.

 What are the effects of counter-urbanisation on the rural settlements involved?

6 Tectonic hazards & rocks

This chapter deals with the causes and effects of tectonic hazards. It then covers rock types.

Tectonic hazards

Tectonic hazards are caused by processes found in the Earth's crust. The main hazards are volcanic eruptions and earthquakes.

The structure of the Earth

The Earth is made up of various layers, as shown in the diagram below:

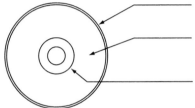

Crust – a relatively thin layer of solid rock.

Mantle – a semi-molten layer of rock to a depth of around 2900 km. May reach 5000°C.

Core – This is made of iron and nickel. The outer layer is semi-molten, but the inner core is solid.

The crust is divided into	plates – massive 'jigsaw pieces' of rock.
The edges of plates are called	plate boundaries.
Examples of plates are	North American plate, Pacific plate.
Britain is on	the Eurasian plate.
The plates move	very slowly – at about the rate your fingernails grow!

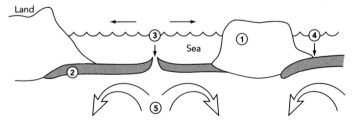

① Continental crust – up to 65 km thick.

② Oceanic crust – 6 to 10 km thick.

③ Mid-ocean ridge

④ Deep ocean trench

⑤ Convection currents in the mantle

Check whether you will be expected to label a map of the world's plates in the exam. If you are, practise with an atlas.

Types of plate boundaries

 Destructive boundaries
are where

	two plates which are moving towards each other meet.

For example,

the Pacific and South American plate boundary on the west edge of South America.

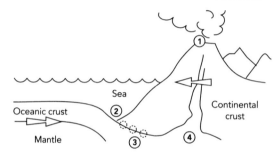

① Violent volcanic eruptions

② Deep ocean trench

③ Earthquakes

④ Oceanic plate melts then rises to the surface as an active volcano

 If the two plates are both continental,

the crust is not destroyed, so they collide and the old sea floor sediment forms fold mountains.

For example,

the Himalayas (Indo-Australian and Eurasian plates meet).

Constructive boundaries are

the border between two plates which are moving away from each other.

For example,

the Mid-Atlantic ridge – where the North American and Eurasian plates move apart.

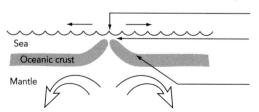

Mid-ocean ridge

Relatively gentle volcanic eruptions and earthquakes

New sea floor or volcanic islands created

| **Conservative boundaries** are | where two plates slide slowly past each other, side by side. |
| For example, | the San Andreas fault, California – where the Pacific and North American plates meet. |

Tip

There are different names for the same types of plate boundary:

Destructive boundary = convergent, trench or compressional.

Constructive boundary = divergent, ridge or tensional.

Conservative boundary = transform.

If you have used a different term than on the diagrams above, that is fine – stick to what you have been taught, because that is what is likely to be used on your exam paper.

Structure of volcanoes

| Most volcanoes are found | on or near plate boundaries (see above). |
| Some are found | over 'hot spots' in the mantle which don't have plate boundaries, eg Hawaii. |

 Label this diagram to show the parts of a volcano by using the key:

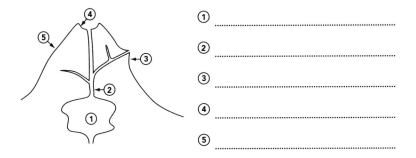

① ....

② ....

③ ....

④ ....

⑤ ....

Lava is

When below ground, it is called

An **eruption** is

An **active volcano** is

Dormant means

Extinct means

molten rock.

magma.

when lava comes out of the volcano.

one which has erupted recently and is likely to erupt again.

the volcano has erupted within the last 2000 years, but is not currently active.

the volcano will not erupt again.

 There are two main types of volcanoes. Complete this table to show the differences between them:

	Cone volcanoes	Shield volcanoes
Lava type	Thick, acid lava	
Lava flow		Hot (1200°C) and runny – spreads over a wide area before cooling
Shape	Steep-sided cone	
Type of eruption		Frequent, relatively gentle – lava and steam
Location	Destructive plate boundaries	
Example		Mauna Loa, Hawaii

A **composite cone** volcano
is made of

layers of ash and lava.

As well as lava flows,
an eruption may involve

falls of ash, nuées ardentes (burning
gas and debris clouds), volcanic
bombs, mudflows and landslides.

Intrusive landforms
are formed by

magma entering the crust but
solidifying before it reaches
the surface.

For example,

dykes and sills.

Earthquakes

An **earthquake** is

a sudden movement of the crust.

Plates do not move smoothly – friction makes the rocks stick, then pressure
builds up, causing an earthquake when the rocks suddenly move.

 The point where an
earthquake starts is called

the **focus**.

The **epicentre** is

the point on the surface directly
above the focus.

The shock waves
will be worst

nearer to the epicentre.

If the earthquake is in or
near the sea, it may cause

a **tsunami** – a giant wave,
eg Chile in 1960.

The **Richter scale**
measures

the strength of an earthquake.

It is numbered

from 0 to 12.

Each point on
the scale is

10 times greater than the point
below.

A reading of 3 is

only recorded by scientific equipment.

A reading of 8 is

a devastating earthquake.

For example,

the 1995 Kobe earthquake, Japan
measured 7.2 on the Richter scale.

Living with tectonic hazards

Effects of tectonic hazards

 Draw a spider diagram to show the 'Benefits and problems of living in an area affected by tectonic hazards'. One to start you off – 'Fertile volcanic soil is good for farming'.

Case study: Kobe Earthquake

Date: 17 January 1995

Severity: 7.2 on the Richter scale

Focus: 20 km deep

Effects:
- over 5000 people killed
- over 30,000 people injured
- over 300,000 people evacuated
- 28% of houses were wholly or partially destroyed
- infrastructure destroyed or disrupted
- many public buildings damaged
- long term effects such as unemployment, loss of business, impacts of bereavement and trauma on people's lives, high costs of rebuilding
- estimated cost of rebuilding and economic loss thought to be up to £90 billion.

 Write out your own case study summary for a volcanic eruption or earthquake you have studied. Remember to include location (plus a map if possible), date, facts about what happened and the short and long term effects.

 Why are the effects of a hazard often more severe in an LEDC than an MEDC?

Response to tectonic hazards

As the human and economic cost of tectonic disasters can be so high, countries with sufficient funds available have put time and money into predicting and preparing for earthquakes and volcanic eruptions.

Volcanoes may be monitored by

- tiltmeters – measure changes in the shape of the ground

- measuring temperature, speed and composition of any existing lava flows

- gas emissions from the ground.

Scientists try to predict earthquakes by

- mapping previous earthquakes to see if there are 'gaps' along the boundary where an earthquake is likely

- seismometers show movements in the crust and may indicate pressure build up

- radon gas emissions from the ground.·

However, this prediction is not accurate enough to be reliable, even in countries such as Japan, which have put a lot of research into it. Even with good prediction, the tectonic forces are too strong to control, though there have been attempts to divert lava flows. To minimise damage, the main option is to prepare well. For example:

- buildings and infrastructure can be built to reduce damage in earthquakes (eg gas pipes with flexible joints, furniture strapped to walls, buildings with steel frames or deep foundations)

- careful planning of cities so that risky areas (eg clay, which turns into mud in an earthquake) are avoided

- people can be trained to deal with a disaster (eg Japan holds 'Earthquake Awareness' days)

- insurance can help reduce economic costs after a disaster.

These measures are expensive, which puts LEDCs at a disadvantage – sometimes the only option may be emergency relief after the event.

Rocks and landscapes

 There are three categories of rocks:

1	**Igneous**	made from cooled magma, eg granite.
2	**Sedimentary**	made from fragments of other rocks or remains of living things compressed into rock, eg sandstone, chalk.
3	**Metamorphic**	rock changed by heat and pressure, eg limestone makes marble, sandstone makes quartzite.

 Draw a spider diagram showing 'Uses of rocks'. Give as many examples of different rock types as you can and next to them, say what they are used for. One to start you off – 'Marble – high quality fireplaces and flooring'.

Geology (rock type) is one of the main factors determining the landscape of an area. The landscape is slowly altered over time by the action of water and ice (see *Chapter 7*), weather and people.

 Weathering is — the disintegration of the surface layer of rock in its original position.

There are three main types of weathering:

1	**Physical** weathering is	when rocks are broken up by physical means. For example, freeze-thaw weathering is when water present in cracks in the rock freezes and expands, shattering the rock.
	Exfoliation is	when the outer layer of rocks in hot dry climates expands more than the inner rock, and eventually peels off.
2	**Chemical** weathering is	when chemicals dissolve the surface of rock, eg rain (slightly acidic) dissolves limestone (**solution**).

3 **Biological**
   weathering is | when living things disintegrate the rock, eg plant roots widen cracks in the rock.

The shape of the land is also changed by:

Erosion – | the wearing away of the land by ice, water or wind.

Mass movement – | the downhill movement of soil, loose stones and rocks due to gravity, without being carried by ice, water or wind.

Weathering and erosion are faster in some areas than in others. Climate and rock type control this. For example, chemical weathering is faster in hot, wet climates. Hard rocks (eg granite) erode more slowly than soft rocks (eg clay). The type and rate of weathering and erosion affect the landscape.

Most rock types have their own distinctive landscape features. For example, in a limestone area, most of the water runs underground, often forming caves (eg the Gaping Gill, Yorkshire Dales, England).

 Name the features on this diagram of an area of limestone scenery:

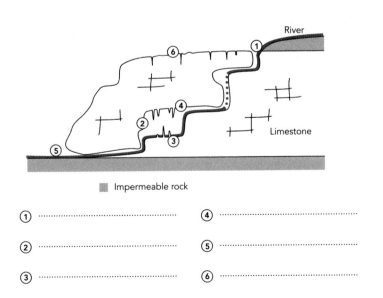

① ....   ④ ....

② ....   ⑤ ....

③ ....   ⑥ ....

7 Water landforms & systems

This chapter covers rivers, coasts and glaciation as well as the management of water supplies for people.

The water cycle

 The majority of the world's water is | salt water (97%).

75% of the fresh water is stored as | ice.

The **hydrosphere** is | the term for everywhere in the world where water is found.

The **water cycle** shows | how water moves around the hydrosphere.

Another name for it is | the **hydrological cycle**.

 Fill in the missing labels on this diagram of the basic water cycle:

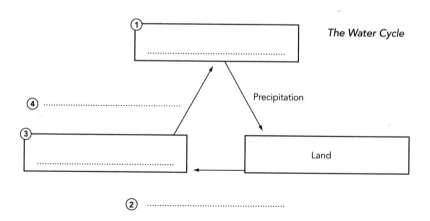

The Water Cycle

① ....

Precipitation

④ ....

③ ....

Land

② ....

Transfer from the land to the sea involves a number of different paths for the water, each with its own name. These are shown on the diagram below:

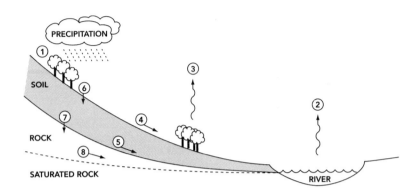

<table>
<tr><td>① Interception</td><td>⑤ Through flow</td></tr>
<tr><td>② Evaporation</td><td>⑥ Infiltration</td></tr>
<tr><td>③ Transpiration</td><td>⑦ Percolation</td></tr>
<tr><td>④ Overland flow</td><td>⑧ Groundwater flow</td></tr>
</table>

 On some rough paper, write out each of the words in the diagram above and jot down what each means. Check your answers in your class notes or the glossary of a textbook.

Water may be stored in	ice (eg glaciers), lakes, ponds, sea, reservoirs, snow or underground.
Overland flow is also called	**surface run-off** (eg streams).
The **water table** is	the level below which the ground is saturated with water.

River landforms and processes

 A **drainage basin** is | the area of land which is drained by one particular river system. It is sometimes called a **river basin** or the **catchment area** for a river.

 Match up the parts of the drainage basin shown on the diagram with their names below:

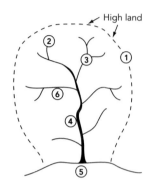

High land

.... Source – the start of a river

.... The main river channel

.... Mouth or estuary – where the river meets the sea

.... Watershed – the edge of the drainage basin

.... Confluence – where two rivers or streams meet

.... Tributary – a smaller river or stream which joins a larger one

Although all rivers are different, the same processes are found, and they tend to produce certain types of landforms at different stages along their path to the sea. A **long profile** is the side view of a river from its start to its end.

The three basic processes in a river system are:

- Erosion | wearing away the banks and any stones the river's carrying.

- Transportation | carrying stones and gravel.
- Deposition | dropping stones and gravel on the river bed.

 Whereabouts on the long profile would you expect to find each of the three processes mentioned above?

The upper river

As the river starts off some distance above sea level, it has power for downward erosion.

 Erosion happens in four main ways:

• **Hydraulic action –**	the force of the water wears away the banks.
• **Corrasion (abrasion) –**	stones the river is carrying bump into the banks and wear them away.
• **Corrosion (solution) –**	acids in the water slowly dissolve the river banks and the stones that the river is carrying (especially limestone).
• **Attrition –**	stones being carried by the river bump into each other and get worn away.
In the upper river, the valley is	V-shaped.
The river may wind between	interlocking spurs.
The water flow is	turbulent (rough), especially after rain.
The river bed is	rocky and uneven – there may be rapids or waterfalls.

An example of a waterfall is High Force, River Tees, Durham, England (20 m high).

 Study the diagrams showing the development of a waterfall over time, then fill in the gaps in the passage overleaf:

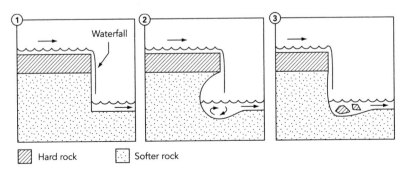

A waterfall usually forms at a break of slope where hard rock overlies

.... The river erodes the softer rock .... than the

harder rock. This creates an .... behind the waterfall. The force

of the water falling, and the movement of rocks in the stream erodes a

.... in the river bed at the base of the waterfall.

Eventually, the hard rock is undercut to the point where

.... and the remains of the overhang is left as

debris on the stream bed. Over time, the waterfall ....

up the valley towards the river's source. It forms a narrow, steep sided

valley, called a .... as it goes.

Transportation

Most material that the river erodes will be carried for at least a short
distance. Transportation happens along most of the length of the river.
The river has more power to carry larger stones near its source or when it is
in flood after heavy rain.

 Name the four types of transportation shown on the diagram:

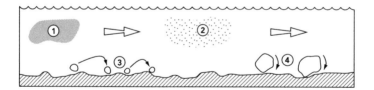

① .... dissolved rock (eg limestone) carried

by water.

② .... small particles carried by the water

(looks muddy).

③ .... pebbles and stones bounce along.

④ .... larger rocks roll along the river bed.

The lower river

Nearer the sea, the river has less power for vertical erosion and often starts to wind across the wide valley. Stones and sand will be deposited at points when the river becomes shallow or slower than usual.

Typical landforms on the lower river:

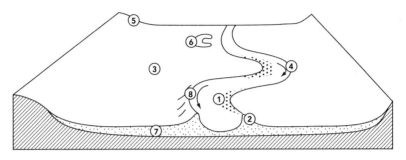

(1) Deposition   (3) Floodplain   (5) Gentle slopes

(2) Levee   (4) Meander   (6) Ox-bow lake

(7) Layers of alluvium (silt) from previous flow

(8) Fastest flow and erosion on outside of bends

 The **floodplain** is | the flat area of land on either side of the river where silt is deposited in times of flood.

It is good for farming because | the **alluvium** (silt) dropped in the flood is fertile.

Levees are | banks on either side of the river which either form naturally through the deposition of silt or are artificial, to prevent floods.

 Explain with the help of a diagram how meanders and ox-bow lakes form. Check your answer with your notes or textbook.

Case study: The Nile Delta, Egypt

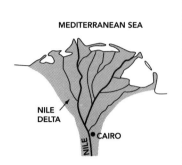

This case study shows the relationship between people and a river landform. A delta is a triangular area of land where the river splits into a number of different channels (distributaries) and drops its sediment on entering the sea. The alluvium dropped at the Nile Delta makes the soil fertile and good for farming. Fish are caught in the river which is well supplied with nutrients.

These advantages make population density high in this area, so people benefit from this landform. However, the situation was altered when the Aswan Dam was built further up the River Nile. The dam reduces sediment flow and controls flooding. This means that the amount of fertile silt being dropped at the delta has decreased, and erosion rates along some parts of the coast are more than one metre per year.

Variations in a river's flow

 Discharge means

the amount of water passing a measuring point on the river over a particular time period.

It is usually measured in

cumecs – cubic metres per second.

A river's discharge varies

over the year and from day to day.

A **hydrograph** is

a graph showing a river's discharge over time.

 Why doesn't all rainfall leave the drainage basin by way of the river channel? Will the same proportion of rainfall be lost all year round?

Look at this storm hydrograph then fill in the blanks in the writing below it:

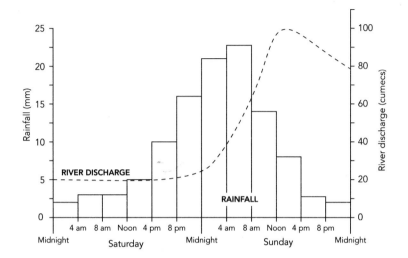

The baseflow for the river is ...... cumecs. Rain starts at .... on Saturday and peaks on Sunday at .... mm. The rising limb of the hydrograph is .... and the lag time is .... hours. This time lag happens because the rainwater takes time to make its way to the river. The peak discharge is .... cumecs. If the channel capacity had been 80 cumecs, the river would have ....

When rain reaches the ground, it will either sink in (**infiltrate**) or flow over the surface (**overland flow**). **Permeable** surfaces allow water to sink in (eg newly dug soil), whilst **impermeable** surfaces do not (eg tarmac). The faster the rainwater gets to the river, the more likely it is to flood. Water will usually travel faster by overland flow than by infiltration and throughflow. Water travels particularly quickly in urban areas because there are many impermeable surfaces, and drains are designed to remove water quickly.

 Draw a spider diagram showing 'Factors which make a river more likely to flood'. One to start you off – 'Steep valley sides'.

101

Case study: The Bangladesh Floods

Much of Bangladesh is situated on the delta of the Ganges and Brahmaputra rivers. The delta provides fertile land for rice-growing, but means that much of the country is very low lying (less than 6 metres above sea level). Flooding is an annual event, but sometimes floods are very severe. For example, in August/September 1988, over 60% of Bangladesh was flooded. This was caused by unusually heavy monsoon rains, and deforestation in the Himalayas may have contributed to the problem by increasing surface run-off. Over 2000 people were killed, buildings and roads were destroyed and crop yields were reduced. Bangladesh is also affected by coastal flooding from storm surges caused by tropical cyclones.

 Write a summary of ways of reducing the risk of flooding (eg building dams, building artificial levees, afforestation, straightening channels). Include a brief case study.

Coastal landforms and processes

The basic processes of erosion, transportation and deposition are the same for coasts as they are for rivers, except that sea water is doing the work instead of river water.

The process of erosion is most obvious in areas where the sea reaches cliffs on a daily basis. As with rivers, softer rocks (eg boulder clay) erode faster than harder rocks (eg granite). Along the coast, an area of harder rock tends to stick out to form a **headland**, whilst softer areas of rock are eroded to make **bays**.

Coastal landforms made by erosion:

① Arch
② Stack
③ Stump
④ Cave
⑤ Wave cut notch
⑥ Wave cut platform

In many areas, the cliffs are protected by a sand or shingle beach. The waves will only erode the cliffs during storm conditions. Beaches are constantly changing.

 Destructive waves are powerful. They move material down the beach towards the sea.

Constructive waves are gentle. They move material up the beach towards the cliffs.

Sand is carried along the coastline by the process of **long shore drift** (see diagram below).

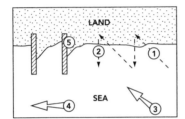

① Incoming wave (swash)

② Outgoing wave (backwash)

③ Prevailing (usual) wind

④ Direction of movement of sand and shingle (long shore drift)

⑤ Material piling up on the windward side of wooden groynes

 Why is the backwash at a different angle to the swash?

Distinctive coastal landforms are also created by deposition.

Draw lines to match up the following landforms with their definitions:

spit •

dune •

bar •

tombolo •

• a small hill of sand at the top of a beach, not usually reached by the tide, which may have vegetation on it.

• a narrow ridge of sand or shingle joining an island to the mainland.

• a long, narrow projection of sand or shingle, joined to the land at one end.

• a narrow area of sand and shingle connecting two headlands across a bay. The area of water it encloses is called a **lagoon**.

Coastal erosion is an ongoing natural process, but it can be a hazard if it occurs rapidly in an area where people live or which is of value for recreation or farming. For example, erosion on the Holderness coast, Humberside, averages at 1.2 metres per year. Various means are available to try to control erosion, such as groynes, sea walls, breakwaters and gabions (metal cages filled with stones). However, these are expensive and will have impacts further down the coastline – if one area is protected, another area may experience increased erosion.

 Write out a summary for a case study of either coastal flooding or coastal management, using your class notes or textbook to help you. Don't forget to include a map and key facts.

Glacial landforms and processes

In the last Ice Age, 30% of the world was covered by ice, including an ice sheet over northern Britain, which extended as far south as the Midlands. During this time, glaciers shaped much of our upland landscape, and the deposition of material they eroded affected a much wider area. Now only a few areas of the world (eg Antarctica) have ice sheets, whilst valley glaciers can be seen in areas such as the Alps.

The diagram below shows a glacier as a system:

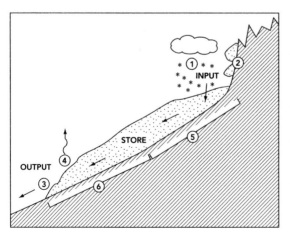

INPUTS
① Fresh snow
② Avalanches

OUTPUTS
③ Meltwater
④ Evaporation

⑤ Zone of accumulation (inputs greater than outputs)

⑥ Zone of ablation (outputs greater than inputs)

Erosion

As with rivers and coasts, the basic processes of erosion, transportation and deposition apply, though the exact methods are a little different.

 Glaciers erode rock in two main ways:

1 **Plucking** –	ice freezes onto rock, then pulls chunks out as it moves on.
2 **Abrasion** –	rocks being carried by the glacier wear away the valley floor and sides as they pass over.

Freeze-thaw weathering also occurs on the valley sides.

Erosion by ice in upland areas forms a series of distinctive landforms. In Britain, these can be seen in areas such as the Lake District and Snowdonia.

 Name the landforms on the sketch below:

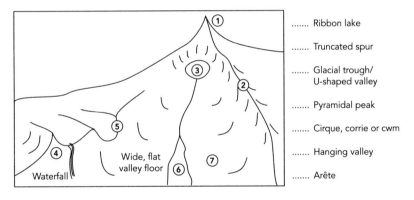

....	Ribbon lake
....	Truncated spur
....	Glacial trough/ U-shaped valley
....	Pyramidal peak
....	Cirque, corrie or cwm
....	Hanging valley
....	Arête

Waterfall — Wide, flat valley floor

 For each of the landforms on the diagram, write one sentence to explain how they were formed. Check your answers with your class notes or your teacher.

Transportation and deposition

 Moraine is

Moraine is	the glacier's load which is eroded from the valley, carried along and later deposited.
It mainly consists of	angular pieces of rock.

It is deposited when	temperatures rise and the glacier starts to melt.
When dropped, it makes	uneven mounds or ridges.

Moraines have different names according to whereabouts in the glacier (and then the valley) they are found.

 Match up these moraine names with their positions:

Lateral •   • in the middle.

Medial •   • over the valley floor.

Terminal •   • across the valley at points where the glacier was stationary during its retreat.

Recessional •   • at the sides of the valley.

Ground •   • across the valley at the furthest point the glacier reached.

Till or boulder clay is	another name for ground moraine.
Drumlins are	mounds of boulder clay deposited by the glacier. They have a streamlined shape due to ice moving over them.

Water and people

Water supply

There are various sources of fresh water:

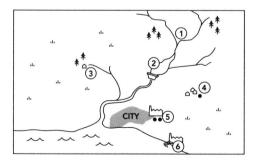

1. Rivers, lakes and streams
2. Reservoirs
3. Spring
4. Well
5. Boreholes
6. Desalination of sea water

In which circumstances might desalination be used?

An **aquifer** is	a layer of permeable rock (eg chalk) under the ground, which is saturated with water.
The water is held there by	impermeable rocks (eg clay) above and below.
The water can be reached by	digging wells and drilling boreholes.
It may also appear naturally at	springs.
An area in the UK which gets water from boreholes is	London.
Water supply in Britain is	uneven – the west receives more rain than the east, but much demand for water is in the east.
This means that	some of the rainfall in the west has to be stored then piped to the east.
The water is stored in	**reservoirs**.
These are	artificial lakes made by damming a river.
For example,	Ladybower Reservoir, Peak District, UK.
A reservoir can also provide	space for sports and leisure.
UK water quality is monitored by	regional water companies and the National Rivers Authority.
A **drought** is	a prolonged, continuous period of dry weather.

Some areas of the world experience severe water shortages due to drought (eg the Sahel region of Africa). These areas also lack adequate equipment for water supply (eg pumps and wells). Britain also experiences short periods of drought which may cause water shortages leading to measures such as hosepipe bans.

Case study:
Water management in an LEDC
– the Aswan Dam, Egypt

The Aswan Dam was finished in 1970. It was built across the River Nile in southern Egypt and it created Lake Nasser (see map). The aim of the dam was to make hydro-electric power and to provide an increased and dependable water supply to grow food. Water outlet from the reservoir is controlled. This means that the annual Nile flood no longer occurs. Less silt is travelling down the river as some of it is trapped behind the dam.

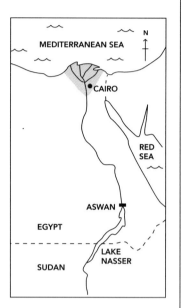

Advantages of the dam
- water provided for irrigation
- constant supply of water
- makes electricity
- enables fishing on the reservoir
- less damage from floods
- easier for ships to navigate down the river without floods

Disadvantages of the dam
- no floods mean that fertile silt is no longer deposited on farm land
- organisms which caused disease (eg bilharzia) no longer washed away in floods
- Nile delta is eroding due to reduced silt supply
- evaporation on irrigated fields causes salt build-up, which is bad for crops
- fewer nutrients in the river after the dam, so fish supplies are declining.

 Think of one group of people in Egypt who have benefited from the building of the dam and one group which has lost out.

 Use your class notes or textbook to make a revision summary for one case study of water pollution. Suitable examples would include pollution in the North Sea, pollution in the River Rhine, or the effects of acid rain in Europe (see pages 51 and 52).

8 Weather & climate

This chapter summarises atmospheric processes and how they affect people.

 Meteorology is

Weather means

Climate is

the study of weather and climate.

the day to day condition of the **atmosphere** (air) in a particular place.

the average weather conditions over an area – usually measured over at least 30 years.

Weather

 Draw a spider diagram to show 'The different aspects of weather'. One to start you off – 'Visibility'. In a different colour, add how you would measure or record each aspect (eg visibility – by eye, in metres or km).

In Britain, information about the weather in different areas is collected by weather stations (staffed and automatic) on land and at sea, planes, radar and satellite. In the exam, you may be asked to describe the weather from a satellite image.

 Find a satellite photo in your class notes or a textbook and practise describing what the weather is like. Remember that water is probably shown in black and clouds will be white. Also look out for satellite images or radar images (for rain) on the TV or newspaper weather forecasts.

The Meteorological Office at Bracknell, UK, collects the data together to make forecasts of the weather. Current weather and forecasts are often shown on synoptic charts (weather maps). These use special symbols to show weather information. Usually there will be a key on the map (check this with your teacher).

 Describe the weather shown by this symbol:

10   1001

Draw the symbol for the following weather conditions:

- southerly wind at 5 knots
- 6/8 cloud cover
- temperature 1°C
- pressure 1010 millibars
- mist

Precipitation

 Precipitation means | moisture travelling from the atmosphere to the Earth's surface – rain, snow, hail, sleet, etc.

Water enters the atmosphere by evaporation. It is carried as invisible water vapour until it is forced to rise (for various reasons), causing cooling and condensing. Cold air can hold less water vapour than warm air. Further cooling leads to precipitation.

There are three main types of rainfall:

1 **Relief rainfall** – air cools as it is forced to rise over hills.

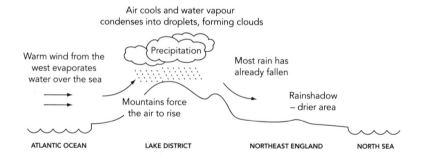

Air cools and water vapour condenses into droplets, forming clouds

Precipitation

Warm wind from the west evaporates water over the sea

Most rain has already fallen

Mountains force the air to rise

Rainshadow – drier area

ATLANTIC OCEAN   LAKE DISTRICT   NORTHEAST ENGLAND   NORTH SEA

A **rainshadow** is | an area with a relatively low rainfall on the leeward side of an area of hills or mountains.

2 **Convectional rainfall** – hot air cools as it rises.

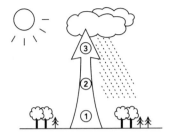

(1) Hot, sunny day
– air near the ground is heated

(2) Warm air rises

(3) It cools and condenses,
leading to rain

Convectional rain is | generally heavy – often a
thunderstorm.

3 **Frontal rainfall** – warm air cools as it rises over a cold air mass.

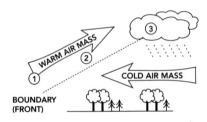

(1) Warm air mass is forced to
rise over the cold air mass

(2) It cools as it rises and
condensation happens

(3) Rain

Frontal rainfall is found in | a depression (see page 113).

Temperature

The temperature range of a place will be influenced by factors such as its
latitude and altitude (see Climate, page 114). On a day to day basis
temperature will be affected by the weather systems present in the area. In
Britain, the temperature, and weather generally, is affected by the movement
of air masses. On a weather map, an isotherm is a line of equal temperature.

 An **air mass** is | a moving quantity of air which has a
particular characteristic.

Polar air is | cold – from the north.

Tropical air is | warm – from the south.

Maritime air is | damp – it carries water evaporated
from the sea.

Continental air is | dry – it has travelled over land, so less
water was available to be evaporated.

A **front** is | the boundary between air masses.

Put these labels on the map of Britain below:

- polar maritime
- tropical continental
- warm and damp
- warm and dry
- tropical maritime
- polar continental
- cold and dry in winter
- cold and damp

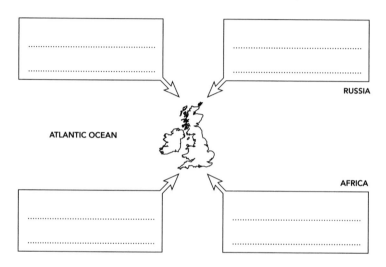

RUSSIA

ATLANTIC OCEAN

AFRICA

Wind

 Air **pressure** means | the force of the air pressing down on a particular place.

It is measured in | **millibars** (eg 1000 mb).

Isobars are | lines of equal pressure on a weather map (like contours on a relief map).

Tip

The numbering on isobars usually goes up in steps of 4, eg 1000, 1004, 1008.

Wind blows from | areas of higher pressure to areas of lower pressure.

Draw in the wind direction on these diagrams:

a

| | | | |
1000 996 992 988

b

——— 988 ———

——— 992 ———

——— 996 ———

——— 1000 ———

| The closer the isobars are, | the stronger the wind. |
| Prevailing wind means | the usual wind direction in a place (from the west in the UK). |

However, in real life, isobars tend to form circular patterns – these are pressure systems.

Pressure systems

There are two major pressure systems – the **depression** and the **anticyclone**.

 Complete this table to show the differences between the two:

	Depression	Anticyclone
Diagram	988 984 980 976 cold warm	1024 1020 1016
Pressure		High
Wind direction	Anticlockwise and inwards	
Wind speed		Usually gentle or calm
Weather	Varies with the passage of the depression – changeable. Cloud. Areas of rain. Warmer in warm sector.	Winter: Summer:

Check with your teacher whether you need to know the detailed sequence of changes associated with the passage of a depression. If you do, write it out in note form. Use a weather map from a newspaper to practise explaining how the weather at one place will change as the depression moves over.

Climate

Climates can be described and explained on a small or large scale.

Microclimates

 A **microclimate** is | the average weather of a particular small area. For example, a wood will have its own microclimate, as will the area round a lake.

 Draw a spider diagram to show 'Factors which will influence the microclimate of a particular area'. One to start you off – 'Aspect' (which direction it faces).

Cities have their own microclimates:

Temperatures in cities tend to be	warmer than the surrounding rural area – often by 2 to 4°C.
This is especially noticeable	during the night.
It is called	the **urban 'heat island'**.
It is caused by	• heating systems in buildings
	• bricks and concrete absorbing the sun's heat in the day and releasing it at night
	• pollution trapping heat.
Windspeed in cities is	generally reduced due to the friction from buildings, but there may be turbulence.
Wind direction is	more varied as it moves round buildings.
Cloud cover is	increased, and there may be smog.

Precipitation is | more intense, but snow will melt more quickly than in rural areas.

Climates on a wider scale

The climate of an area is influenced by the following factors:

- Latitude – | places nearer to the equator are warmer because the sun is more directly overhead.

- Distance from the sea – | the sea has a modifying influence on climate – places near the sea are cooler in summer and warmer in winter, compared to inland.

- Altitude – | places that are higher above sea level (eg tops of hills) tend to be colder, windier and get more rain.

- Prevailing winds – | if winds come from the sea, they will be damp and generally speaking mild in winter, cool in summer. If winds come from over the land, they will be dry and cold in winter and hot in summer.

Britain's climate

Britain's climate can be described as **temperate maritime**. Temperate means that it is not extreme, and maritime means it is influenced by the sea. However, different areas of Britain have their own individual climates, and the weather involved can vary considerably from year to year!

 Draw lines to match up the areas of Britain with their climates:

North West •   • Hot summers, cold winters, dry

South East •   • Cool summers, cold winters, fairly dry

North East •   • Warm summers, mild winters, wet

South West •   • Cool summers, fairly mild winters, wet

 Find a map showing average annual rainfall and winter/summer temperature patterns for Britain in an atlas or textbook. Write a few sentences describing the patterns you can see on each map. Don't forget to quote some figures.

The west side of Britain is warmed by	the North Atlantic Drift.
This is a warm water current which	travels across the Atlantic from the Gulf of Mexico.
The west side of Britain is wetter due to	relief rainfall occurring when the moist prevailing westerly wind meets high land, eg the Lake District.

The climate of a place is often shown on a double axis climate graph.

| Temperature is shown by | the line. |
| Rainfall is shown by | the bars. |

Make sure you're reading off the correct axis – check the labels!

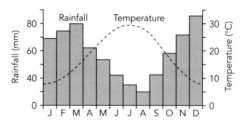

Major climates of the world

There are various major climatic regions in the world, such as the Equatorial climate, Arctic climate, and hot desert. These are linked with biomes (see page 42). The type of climate in each region is determined by the factors on page 115.

Case study: The hot desert climate

Location: Mostly between 10° to 30° north or south of the equator.

Examples: Sahara desert, Africa; Atacama desert, South America.

Temperatures: Generally high. May reach 50°C

Temperature variation: Wide variation between day and night due to lack of cloud.

Precipitation: Dry due to prevailing winds travelling over land, and location in the rainshadow of mountains. All deserts get rain at some point, often heavy convectional rain, which may cause flash floods.

Weather hazards

Most extreme weather conditions can be hazardous, eg blizzards, gales and heat waves.

Hurricanes

A **hurricane** is	a very intense low pressure system with winds of over 117 km per hour and heavy rain.
They form over	warm oceans (warmer than 27°C).
They are called hurricanes	if they form over the west Atlantic Ocean and Gulf of Mexico.
Typhoon is the name for	hurricanes in southeast Asia.
The **Saffir-Simpson scale** is	a measurement of the severity of a hurricane.

 Draw a spider diagram to show the types of damage that a hurricane can cause.

The movement and growth of hurricanes can be predicted with a reasonable level of accuracy using satellite imagery, special planes and radar equipment. Countries like the USA have put a lot of money and research into this so that they can warn and evacuate people in places such as Florida which may be at risk. LEDCs may suffer more damage through having less developed prediction and communication systems.

Case study: Hurricane Georges, Caribbean, September 1998

Hurricane Georges started off as a depression in the Eastern Atlantic Ocean. As it moved west, it became more intense with faster winds and gained hurricane status on 17 September. Whilst still out at sea, Georges had winds of up to 150 miles per hour (4 on the Saffir-Simpson scale). Georges travelled near or over various Caribbean islands, then hit Puerto Rico on 21 September with winds of 115 mph. The next day it was even stronger when it reached the Dominican Republic. It weakened when travelling over Cuba, then grew stronger when going across the sea towards the USA, reaching the Florida Keys on 25 September. It carried on in a north-westerly direction and died away after travelling over land in Mississippi. Over 300 people were killed, many of these in the poor areas of the Dominican Republic which was hit by floods and storm surges.

Answers

The following answers are to the questions denoted by and

Page 7
Road bridge; factory; cliff; railway station; orchard

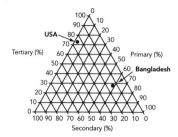

From	To	Direction
P	ⴖ	N
P	▲	SW
●	ⴖ	NW or NNW
▲	ⵌ	NE
PH	TH	SE or ESE

Page 8
B = 0250; C = 0150; D = 0151
▲ = 012500; ■ = 017509

Page 9
$^1/_2$ km; 2 km; 6 km
$^1/_4$ km; 2 $^1/_2$ km
8.7 cm = 4.35 km

Page 10
1 km²; $^1/_4$ km²; $^1/_2$ km²

Page 11
 -5.1 to -10 %

A broken ring around the inner city. Overall pattern shows less migration from the suburbs and more from the centre.

Page 12
Hilly; north; 278 m; uneven; stepped; plateau
1 km; south west; A17

Page 13
Inner city; $^3/_4$ km; west; 11; linear

Page 14
50; 5

Page 15

Page 16
1 day
6 amenities

Page 17

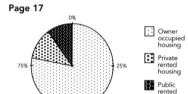

Page 19
Primary 2%; Secondary 28%; Tertiary 70%

Page 21
Very dependent on the price of that product – if the world price falls and/or demand for the product drops, the country's income will drop and people will lose their jobs.

Page 22
Amount of rainfall; relief; temperatures + length of growing season; soil type; aspect (direction in which the farm faces); farm size; capital available; type and size of market; background and training of farmer; government policy; technology available; transport available.

Page 23
Hill farming affects the local environment by encouraging stone walls to be repaired and new lambing sheds built. Arable farming causes trees and hedges to be removed and fields made bigger so that large machinery can be used.

Page 24
Hedgerow removal; genetically modified crops (1998+); introduction of artificial pesticides, herbicides (kill weeds) and fertilisers; vehicles such as tractors replace horses; fewer workers needed; larger fields in arable areas; more indoor rearing of animals and birds; farms get larger;

new crops, eg oil seed rape; drainage of marshy land; some streams straightened and made into ditches; new farm buildings; new government policies, eg set-aside.

Page 25

Why: To make larger fields to grow more crops and to make room for larger machines; some farmers worried that hedges harboured pests and took nutrients from the fields; hedges took time and money to maintain. Problems: windbreaks removed – soil may erode in strong wind or heavy rain; decline in wildlife which used hedges for food and shelter; some people think countryside looks less attractive.

Disadvantage of green revolution: increased reliance on expensive western technology, eg machines, fertilisers; increased rich-poor gap as some cannot afford new technology; debt; health hazard from poor safety procedures when using pesticides; increased rural unemployment due to mechanisation.

Page 27

Suitable workforce nearby; good road access for workforce; sheltered area of coast for dock, with flat land near; originally raw materials nearby, though now often imported; access to market for ships; associated services and manufacturers nearby.

Page 28

1967 – coastal and inland; 1997 – all coastal; clusters stay in some areas of the country, eg south Wales; (note: you could count the numbers in different areas).

Page 29

World recession in 1970s; competition from abroad; UK's raw materials, eg iron ore became more expensive as supplies reduced; changing markets, eg less demand for ships, reduced demand for steel as new metals used; many factories were built in Victorian times and were becoming inefficient, inner city locations were not suitable.

Page 31

Glasgow ⟶ Edinburgh = 'Silicon Glen'; South Wales; Bristol ⟶ London = M4 corridor; Cambridge ⟶ London = M11 corridor.

R&D needs to be near good communications and company headquarters, often in south east; prestige locations in south east for R&D, eg Cambridge; cheaper land and workforce in north for manufacturing, more government incentives in north.

Page 32

Fast road access to London (company headquarters) and Midlands; access to airports such as Stansted (30 minutes) and Heathrow (1 to 2 hours); room for expansion; cheaper land on edge of city; links with university; prestige from 'Cambridge' name; 'greenfield' site means development can be tailored to individual needs.

Page 33

(clockwise) People move into the area, unemployment lower; area becomes more wealthy, local shops and services benefit; council invests in infrastructure and services.

If cycle continues for a long time, area may become over-developed leading to problems with congestion and high house prices. Some firms may start to move away.

page 34

a   Advantages – cheaper production costs (labour, land) = more profit; may be incentives, eg export processing zones. Disadvantages – may be higher transport costs.

b   Advantages – jobs provided; may invest in infrastructure; may enhance reputation. Disadvantages – jobs not well paid; much of the profit goes back to the company's own country; may have environmental consequences.

page 36

1 Lake District, 2 Peak District, 3 Snowdonia, 4 Dartmoor, 5 Norfolk Broads, 6 Yorkshire Dales.

page 37

Advantages – attracts people to the area; employment provided for local people; money into local economy from food, souvenirs, etc; may help protect local wildlife. Disadvantages – much of the profit may go to tour companies based in MEDCs – even food may be imported, so local economy may not profit (especially true of 'all-inclusive' resorts); jobs are low paid and may be seasonal; tourists' dress and behaviour may cause offence to locals; land taken up for hotels; industry may decline as fashions for holidays change.

page 39

Inputs; water; flow or transfer; store; outputs; cycles; the water cycle; liquid; gas. (clockwise from top) carbon dioxide; photosynthesis; litter; decomposition; soil store; roots; respiration.

page 41

Producer – marsh plant; Primary consumer – Eats plants, snail; Secondary consumer – eats herbivores, eg owl; Herbivore – eats plants, fish; Carnivore – eats meat, heron; Decomposers – feeds off dead matter, worm.
Abiotic – sun, air, mud, water – rest are biotic.

Answers

page 42

	Taiga (coniferous forest)
	Savanna
	Hot desert
	Rainforest

page 44

(from top) emergent; canopy; sub-canopy/under canopy; shrub layer; ground layer.

page 45

Leaves designed to shed water, eg drip tips; hot all year round so no one season for shedding leaves; competition for sunlight leads to tall trees.

Grazing land for cattle; space for mining; space for roads and railways; space for farming, eg plantations; land for building houses; logging to get wood especially valuable hardwoods, eg mahogany.

page 46

Extinction of plants which may have medicinal value; loss of habitat; species become extinct; loss of biodiversity and gene pool; indigenous people forced to move away; soil erosion; flooding.

a   Flooding – trees removed ➝ no leaves to break fall of rain ➝ soil surface becomes compacted or eroded ➝ less infiltration and more overland flow ➝ floods. Also eroded soil fills river channels and no tree roots to take up water.

b   Settlers from other areas bring in western diseases which indigenous peoples aren't used to, so may die from.

c   Vegetation gone ➝ less evapotranspiration ➝ drier air.

page 48

Long roots to reach water; some plants grow up quickly in the rainy season then produce seeds and die back in the dry season; some plants store water, eg baobab tree; some have thorny or waxy leaves to reduce evapotranspiration in dry season.

page 49

Overcultivation; overgrazing; climate change due to global warming?; overpopulation; deforestation for firewood.

page 50

Oil; coal; gas; nuclear; hydroelectric power; wave; wind; tidal; solar; biomass (plant waste); animal waste (burning manure); wood; geothermal; rubbish; wind.

page 51

Dominated by fossil fuels, eg oil 33%, coal 25%, some nuclear, small amount of renewables. An LEDC is likely to use more wood and less nuclear, though the exact energy mix would vary according to the natural resources which the country had.

1   Sulphur dioxide; nitrogen oxide; coal; oil; natural gas; cars; factories and power stations.
2   condensation
3   wind
4   2.4 Scotland

Page 52

Buildings and statues eroded (especially limestone); leaves discolour and fall off; trees die; nutrients leached from soils; heavy metals released from soil into water; life in lakes dies.

page 53

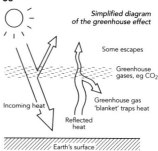

Simplified diagram of the greenhouse effect

Farming areas shift position as climate alters; sea water expands and ice melts causing flooding; refugees move from flooded areas; more extreme weather events, eg hurricanes; increased coastal erosion; more favourable climates in some areas.

Page 56

Sparse: Himalayas (Asia); Greenland
Dense: W. Europe; E. China

Climate; availability of resources; fertility of land; relief; ease of access; natural vegetation; availability of jobs; food and water supply.

Increase; decrease; be stable (stay the same).

page 57

High; slow; death rate; birth rate; growth; decreased; stabilised; low; slow.

page 58

Encouraging larger families by adverts or tax incentives (eg ancient Rome); encouraging smaller families by free and legal birth control, improving education or even passing laws (eg China's one child policy).

page 59

2%; 12.2% (add male and female); more even – less triangular; more equal birth and death rates – low population growth.

page 60

UK total dependants = 32.1% (add young and old) India elderly = 4.1%; children = 39.5%; total = 43.6%